Schools Council/Loughborough University of Techn Engineering Science Project

This Project began in 1970 and continued for four years at the Engineering Science Development Unit (ESDU), University of Technology, Loughborough, with financial support from the Schools Council and from Loughborough University. The project evolved from earlier Joint Matriculation Board efforts to provide schools with Advanced Level texts in Engineering Science and has extended the original scheme to include resource material for students and teachers.

The project's primary aim has been to produce stimulating and enjoyable text books by adopting an engineering approach to the treatment of the physical science included in a wide range of A Level courses. Our educational objectives include the development of abilities in communication, synthesis, design, evaluation and decision-making in addition to the more traditional abilities of analysis and comprehension.

An attempt has been made in each of the texts to treat the science in the context of the ways in which it is applied to the solution of practical problems. Of necessity, therefore, the familiar concepts of physics are considered alongside ideas of optimisation, of engineering judgement and decision, and of economic and social resources and restraints. Science itself is seen as both a powerful tool and an inflexible limiting agent.

Each of the student texts develops a coherent section of physical science by showing its importance and relevance to an engineering situation or situations. In order to make this development more readable a dual treatment has been adopted throughout the series. A complete list of the texts and supporting material is given on the back cover.

Of necessity a number of physical concepts have relevance to more than one text. Where they were felt to be of fundamental importance and central to the engineering situation under consideration, parallel treatments have been given. In other cases cross-referencing has been used, and the books are referred to simply by their titles.

The emphasis has been on the broad development of ideas rather than on rigorous and deep mathematical treatments. Nevertheless, opportunities have been taken to demonstrate the usefulness, power and elegance of mathematics and to cater for students with fair mathematical ability.

The whole series has been subjected to trials in a number of schools and colleges, and the Engineering Science Development Unit has been advised throughout its existence by a Steering Committee composed of industrial and university engineers, teachers in secondary and further education and Schools Council officers. The trial schools and colleges, together with the names of past and present members of the Steering Committee, are listed on page vi.

There has been no attempt to produce a set of complete teaching instructions, nor on the other hand has the intention been to obviate the need for classroom teaching by providing a complete student's study guide. It is hoped that teachers will be able to use the books to supplement their own individual styles of teaching and that students working on their own will find them readable and useful.

D. T. Kelly,
Department of Education (ESDU),
University of Technology,
Loughborough.

The Engineering Science Development Unit

Director
Professor L. M. Cantor, M.A.

Leader
D. T. Kelly, B.Sc.

Staff
G. W. Pilliner, B.Sc. A.R.I.C.
H. Cole, B.Sc.(Eng.)

Authors
M. D. Clark B.Sc.(Eng.)   Elgin Technical College
A. Clarke   CEGB North Western Region
A. Craven B.Sc.   City of Leeds School
R. D. Harrison B.Sc., M.Inst. P.   Open University
J. E. Jones M.A.   Huddersfield New College
C. Williams M.A., Dip. Ed.   Ellesmere Port Grammar School

L. M. Cantor
H. Cole
D. T. Kelly
G. W. Pilliner

Editors
H. Cole
D. T. Kelly
G. W. Pilliner

Schools Council/Loughborough University of Technology
Engineering Science Project

Tribology

Published for the Schools Council by Macmillan Education

© Schools Council Publications 1974

All rights reserved. No part
of this publication may be reproduced
or transmitted, in any form or
by any means, without permission.

First published 1974

Published by
MACMILLAN EDUCATION LIMITED
London and Basingstoke

Associated companies and representatives
throughout the world

Printed and bound in Great Britain at
The Pitman Press, Bath

Contents

How to use this book — viii

1 Tribology and industry — 1

Text A Electrical power generation; Friction: a necessary evil; Coal handling; Ash handling; Planned maintenance

Text B The generation of electricity from coal; The force of friction; Coefficient of friction; Angle of friction

2 Bearings — 11

Text A Plain bearings; Turbine generator main bearing; Thrust bearings; Roller bearings; Linear bearings; Choice of bearing

Text B Friction and wear in plain bearings; Bearing material; The tilting pad thrust bearing; Rolling friction; Types of rolling bearings

3 Lubrication systems — 18

Text A Oil lubricators; The lubrication of gears; Grease lubrication; Turbine generator lubricating system; Bearing seals

Text B Lubricants; Lubricating oils; Alternative lubricants; Surface tension; Capillary action

4 The nature of surfaces — 30

Text A Rough or smooth?

Text B Centre-line average; Hardness

5 The selection of lubricants — 33

Text A Viscosity; The units of viscosity; The measurement of viscosity (Viscometry); Capillary viscometers; Efflux viscometers; Degradation; Consistency; The choice of lubricants

Text B Viscosity; The concentric oil bearing; Viscosity Index; Hydrodynamic or fluid lubrication; Boundary lubrication; Extreme pressure lubrication; Hydrostatic lubrication; Gas lubrication

6 Friction devices — 43

Text A Conveyor belts; Clutches; Brakes; Internal expanding brakes; Disc brakes; Fasteners; Clamps and cotters

Text B The conveyor belt; Belt drives; Friction clutch; Action of simple external brake; Action of a simple drum brake; Disc brakes; Friction in screw threads; Efficiency of screw threads; Jamming

Index — 59

Examination index — 61

Cover: The giant SKF spherical roller bearing that was installed in the Cumberland Basin Swing Bridge in Bristol. The bridge is a balanced cantilever, pivoted at its centre, and allows ships to enter the City Docks at Bristol. The bearing has to support a moving weight of approximately 9 000 kN. Photograph reproduced by courtesy of the Skefko Ball Bearing Company Limited of Luton.

Acknowledgements

We are most grateful to the Central Electricity Generating Board for their continuing support and co-operation, in particular for allowing us to use statistics connected with the electricity supply industry, and for permission to reproduce Fig. 1. We have also made use of CEGB material in preparing the diagram and description on pages 2 and 3.

We are indebted to Messrs International Combustion Ltd of Derby for permitting us to reproduce Fig. 3; to Messrs GEC Turbine Generators Ltd, of Rugby, for the use of the diagram of the turbine generator bearing in Fig. 14; and to the Skefko Ball Bearing Company Ltd, of Luton, for allowing us to use the photograph on the front cover.

We would like to thank everyone who has made useful suggestions; in particular the assistance of the members of the Steering Committee, of colleagues within the university, and teachers and students in the trial schools and colleges has been invaluable. It must be stressed, however, that no blame can be attached to them for errors in formulation or factual matter. The Unit is fully responsible for any such errors that may have eluded our various checking procedures.

Miss A. Cartwright and Mrs H. Rankin whose untiring efforts have considerably eased our burden.

Steering Committee

Mr C. E. Aspinall
Mr G. Bleach
Professor J. N. Butters
Dr T. A. Burdett
Professor L. M. Cantor
Professor G. Carter
Mr T. N. Corkhill
Mr M. T. Deere
Mr W. A. G. Easton
Professor H. Edels
Mr N. D. C. Harris
Mr G. B. Harrison
Mr G. Price
Mr M. Sayer
Dr C. Selby
Mr D. T. Kelly (ex officio)

Trial Schools and Colleges

Archbishop Holgate's Grammar School, York
Blackpool Collegiate Grammar School
Bradford Grammar School
Brockenhurst Sixth Form College
Cardinal Hinsley R.C. School, Bradford
Chesterfield College of Technology
City of Leeds School
Foxwood School, Leeds
Harrogate College of Further Education
Hurstpierpoint College
Keresley Newland High School
King Edward VII Upper School, Coalville
Kirkby Stephen Grammar School
Longslade Upper School, Birstall, Leicester
Loughborough Technical College
Stoke-on-Trent Sixth Form College

Preface

Tribology is a relatively new word to describe an old science and technology, namely the understanding of the nature of surfaces and their behaviour in real circumstances. It covers the study of the structure of surfaces, the disposition of atoms, their mutual forces and their interaction with the environment, and, in technological terms, the friction between surfaces, wear processes, adhesion and lubrication.

Friction and wear are necessary and expensive evils of everyday life in a technological society. Overcoming the forces of friction between moving parts in machines requires expenditure of energy, and a consequent reduction in efficiency; wear is a major contributory factor in limiting the life of moving parts. On the other hand, situations such as traction of wheels on roads and railway lines and the mountain climber's boots on rock rely on friction for effective operation. Wear processes are readily seen in everyday life in the loss of material from the soles of shoes, the rubber from tyres and even the skin from hands and feet.

In a society facing the prospect of dwindling sources of energy such as oil and coal and the rapid depletion of natural mineral resources, the minimisation of friction and wear is of paramount importance. In attempts to minimise such effects, the science and technology of lubrication has grown rapidly and we are all aware of the use of oils in motor car engines (particularly when the level is low, and seizure occurs), and the use of talcum powder before wearing rubber gloves to ensure subsequent easy removal from the hands. We are equally aware of the other side of the coin, e.g. the disastrous effects of ice and thin films of water or oil on roads on the automobile accident rate.

This text introduces the reader to the phenomena of friction, wear and lubrication through specific reference to a major technological undertaking (the coal-fired power station) but as must be obvious from our brief discussion, the consequences of these effects are of much more general importance in our lives. Before and during the study of this text try to imagine how different life would be in a world without friction or wear; in many situations that is precisely the idealised aim of the work of the tribologist.

As is true in any engineering activity, however, these aims are compromised by the restraints of cost and resource availability and scientific laws, as well as by limitations of knowledge and understanding; solutions to problems are optimised according both to society's needs and requirements and to what it is prepared to pay.

Professor G. Carter   *University of Salford*

How to use this book

During the course of your studies you will need to use this book in a number of ways. You will want to obtain an overall picture of the subject and you will need more detailed knowledge of certain important topics. There will be facts and laws which you will have to learn and use, particularly in the calculations that are essential to any branch of engineering, and you will want this book to provide these in an understandable form. In addition, there will be occasions when you need a particular piece of information, and when examination time approaches you will want a book that can be used for rapid revision.

The layout of this book is more complex than in the usual textbook in order to fulfil all these requirements more readily. The book is written virtually in two parts. In text A, the upper half of each page, there is an account of the basic scientific and engineering concepts relevant to the study of friction, lubrication and wear; these concepts and their practical realisation are considered in more detail in text B, the lower half of each page.

Your teacher will advise you when to start work on this book and which parts are relevant to the topics he is covering in class. It is suggested that in working through a topic you work through the relevant part of the book twice.

For the first reading, you should read text A only. This will give you an overall picture of the significance of the topic before embarking on the essential detailed work relevant to it. Re-read text A working through the corresponding text B. We have arranged the pages so that the detailed work in text B applies as far as possible to the text A material above it. On some of the pages questions have been included in both parts of the book. Try to find answers to these during the second reading.

An index has been provided to enable the book to be used for reference purposes. There is also an examination index which will help in the revision of sections of the text which must be known in detail to satisfy particular syllabus requirements. You may like to use this examination index to plan your revision time-table.

Text A (the top half of each page)
An account of relevant scientific and engineering concepts.

Text B (the lower half of each page)
Physical laws and relationships and their mathematical derivation, supporting technical information and suggestions for practical work.

1 Tribology and industry

In British industry vast sums of money are spent each year on the repair and replacement of worn parts in all types of plant and equipment. Wear is a major limiting factor in the life of all machines and mechanisms and is due principally to what is called 'friction' between surfaces that rub together. Obviously we will be able to design against wear if we know something about the nature and effects of rubbing surfaces.

In 1966, following a report published by the Department of Education and Science, the name *Tribology* was given to the theory and practice of interacting surfaces in relative motion. The new name was recently introduced, simply because other terms in common use (such as friction, wear and lubrication) refer only to a part of the technology of moving surfaces in contact. Tribology covers *all* aspects of this technology including friction, bearing-design, lubrication systems and wear. It is mainly concerned with machines and mechanisms, but there are also cases where surfaces rub together outside a machine, e.g. tyres on road surfaces and expansion joints in bridges.

In the following sections we shall consider how the study of aspects of tribology can influence the operating efficiency and maintenance of machines and equipment in industry. Tribology will play an important part in any industry that utilises machines of high capital value which must operate efficiently and continuously for many years. One such area is the electrical supply industry and in this book we shall first examine the importance of tribology to a coal-fired power station (see diagrams on pages 2 and 4).

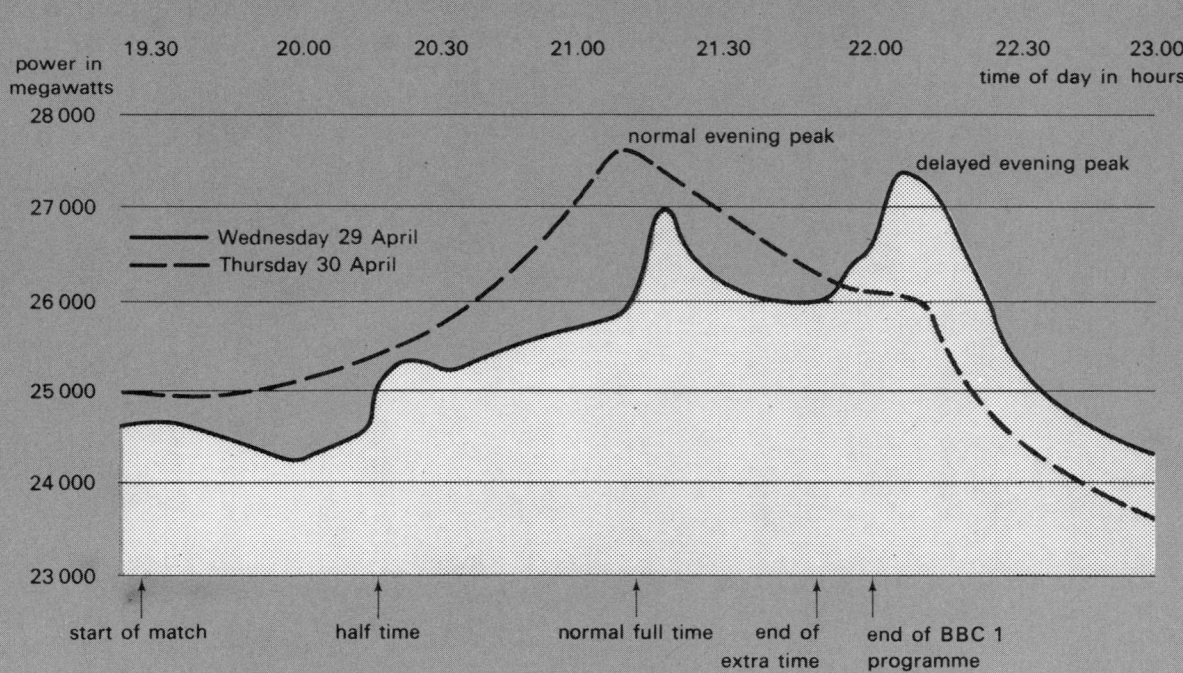

Fig. 1 Power station maintenance engineers must ensure that shut-down plant is well maintained and capable of being run up to full output at short notice, since the demand for electricity is difficult to predict. Television programmes, such as the 'Miss World' contest and some sports programmes have a profound effect on domestic demands for electrical power. The diagram shows the fluctuating demand during the FA Cup Final Replay on Wednesday, 29 April 1970, contrasted with the load curve for the following night. At the end of television coverage viewers all over the country switched on fires, kettles and extra lights.

Electrical power generation

During the period 1960—70 the Central Electricity Generating Board almost doubled the production of electricity in England and Wales from 94 000 million units to over 180 000 million units to meet domestic and industrial demand for electrical power. The generating capacity at the end of 1971 was 54 322 megawatts, which represented an employed capital value of £3 050 m. Most of the electricity sold in 1971 was generated in coal-burning power stations which consumed 61.4×10^9 kg of coal. Oil-burning power stations consumed the equivalent of 23×10^9 kg of coal, and the electricity produced by nuclear stations was equivalent to the consumption of 8×10^9 kg of coal. Coal, then, continues to be the main source of fuel for the production of electrical power.

A power station should show a financial profit and some obligation to deliver the product, electricity, when it is wanted by the customer. In order to achieve these aims the power station must be able to provide power on demand, and, since electricity cannot be stored, the generating plant in a power station may be required to be in full operation in a matter of hours depending on the anticipated variation of electrical load requirements (see Fig. 1). This means, of course, that the equipment must be efficiently maintained by rigorous routine servicing and periodic overhaul. Maintenance must be carefully programmed so that the different parts of the plant are out of service at the most convenient time, usually during the summer when the demand for electricity is low. Ancillary equipment must be overhauled at the same time.

All power stations have a team of maintenance engineers and it is up to them to seek ways and means of prolonging the service life of the machinery, particularly in equipment where the wear rate is rapid or where worn parts are expensive to replace. Maintenance engineers are not only concerned with wear, however; fatigue and corrosion in one form or another also cause failure of equipment in everyday use. (These factors are discussed at greater length in *The Use of Materials*.)

Friction: a necessary evil

Friction is the resistance to motion which is set up when one body slides over another; in

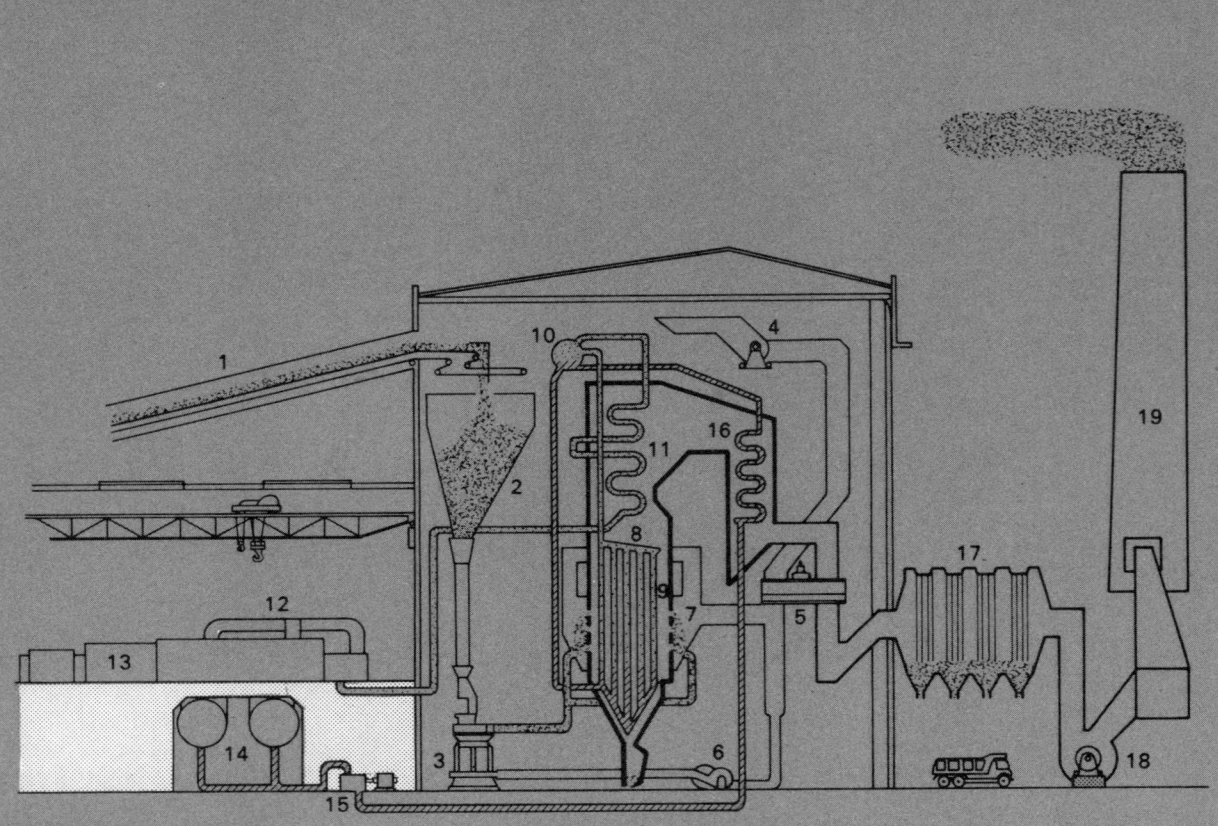

Fig. 2 Simplified section through a coal-fired power station

many instances it is a source of trouble. It means that in every machine some energy must be used to overcome friction at the place where surfaces slide over each other. By using a *lubricant* such as oil or grease between the surfaces, the frictional force can be reduced but never eliminated. So it seems that friction is undesirable, wasteful of energy and costly in terms of the wear it creates and the power expended in machines to overcome it. There are, however, many situations in which friction plays an essential part (in fact you would never be able to walk if there were no frictional forces between your feet and the ground). At the same time the wear created by this useful friction is often a problem. A power station provides many examples of wear associated with useful and necessary friction forces.

In any one day in a large coal-fired power station some 10^7 kg of coal will have to be moved from the coal stock at ground level to the high-level coal bunkers to maintain the flow of coal to the milling plant and hence to the boiler furnace. Much of this bulk handling is done speedily and efficiently by the use of belt conveyors possibly moving 10^6 kg of coal per hour, at speeds of 2 metres per second or greater, up gradients of, perhaps, 1:4. This method of moving coal relies on the action of frictional force between the coal and the conveyor belt surface, usually rubber or plastic, and also between the belt and the electrically-driven drum roller which keeps the endless belt in motion. The tension on the belt must be sufficient to generate the required torque (turning force) to move it, but not so great that electrical power is wasted. Conveyor belts do, of course, wear under arduous conditions like this, but not too rapidly if the coal is delivered on to the belt correctly and the conveyor design is based on narrow belts and high belt speeds.

Belt drives of various kinds are used throughout the station, and, like the conveyor belt, they depend on friction forces. These drives consist of grooved pulleys, one on the motor shaft, the other on the driven shaft (such as the shaft of a fan or compressor), connected together by flexible ropes or belts located in the grooves. Power is transmitted by virtue of the friction force between the belt and the side of the pulley grooves. Belt drives and conveyor belts are discussed in detail in Chapter Six.

The generation of electricity from coal

Figure 2 opposite shows a simplified section through a modern coal-fired power station.

A conveyor belt 1 carries coal from the station coal store and discharges it into a bunker 2. It falls by gravity through a chute into the pulverising mill 3 where it is ground into a fine powder. The mill normally comprises large rollers or steel balls which crush the coal on a slowly rotating grinding table. Warm air is drawn from the roof of the boiler house by a forced-draught fan 4 which sends it through the air-preheater 5. The primary air fan 6 blows some of the hot air through the mill where it dries the coal.

The pulverised coal-and-air mixture passes through ducts to the boiler and enters the boiler furnace at the burners 7 which are so arranged that the pulverised fuel spirals upwards through the furnace 8. The coal-and-air mixture ignites and burns fiercely, heating the water in the boiler tubes 9 and changing it into steam which passes to the steam drum 10 at a high pressure (17×10^6 N m^{-2} in a 500 MW boiler). Further heating in the tubes of the superheater 11 raises the temperature to 568 °C, and the superheated steam passes to the steam turbine 12 driving the generator 13 which generates electrical power. After expansion in the turbine the steam exhausts into the water-cooled condensers 14, and the condensate is returned to the boiler in a continuous cycle by a boiler feed pump 15.

At the top of the boiler the ash and waste gases, (flue gases) resulting from the combustion of the coal, pass downwards through an economiser 16 and give up some of their heat to the boiler feed water in the economiser tubes. The flue gases then pass through the air-preheater 5 which transfers heat from the gases to the air going into the boiler. Over 99 per cent of the dusty pulverised fuel ash which results from burning the very finely ground coal is taken out of the flue gases by an electrostatic precipitator 17. An induced draught fan 18 draws the gases into the chimney 19 which disperses the remainder of the ash so high in the air that it cannot cause smog. Ash collects in hoppers under the precipitators and at the base of the boiler, and is fed into an ash sluice and discharged into wagons or barges for removal.

Figure 4 overleaf shows the flow of air, water and fuel through the main parts of the power station.

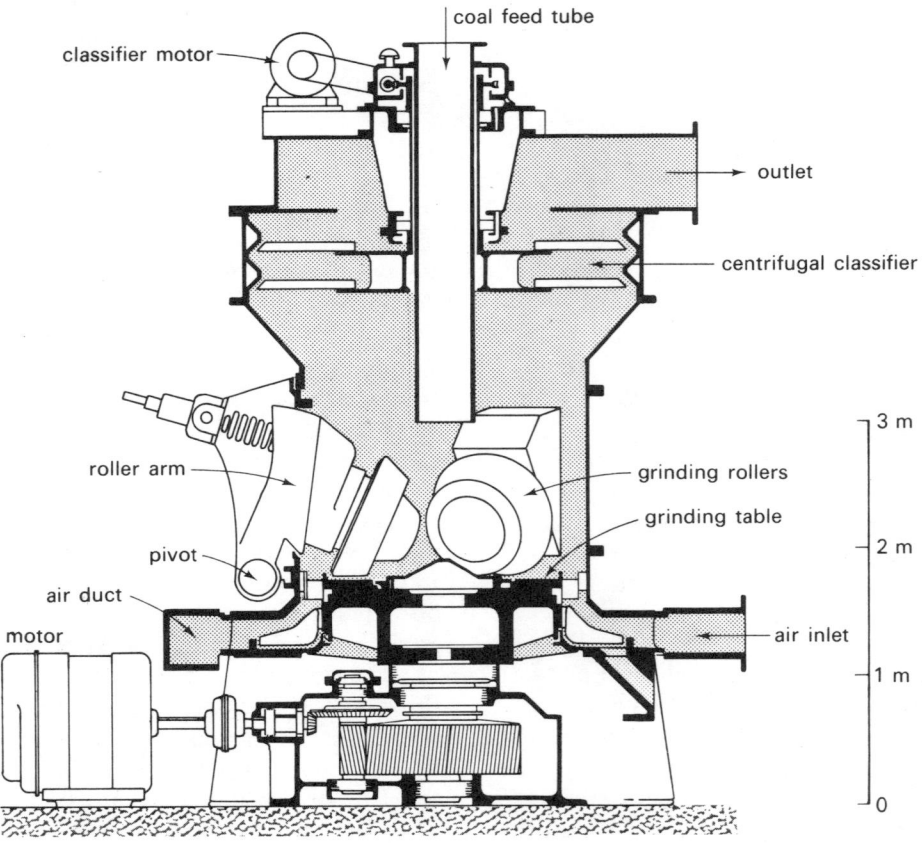

Fig. 3 Section through a large roller-type pulverising mill

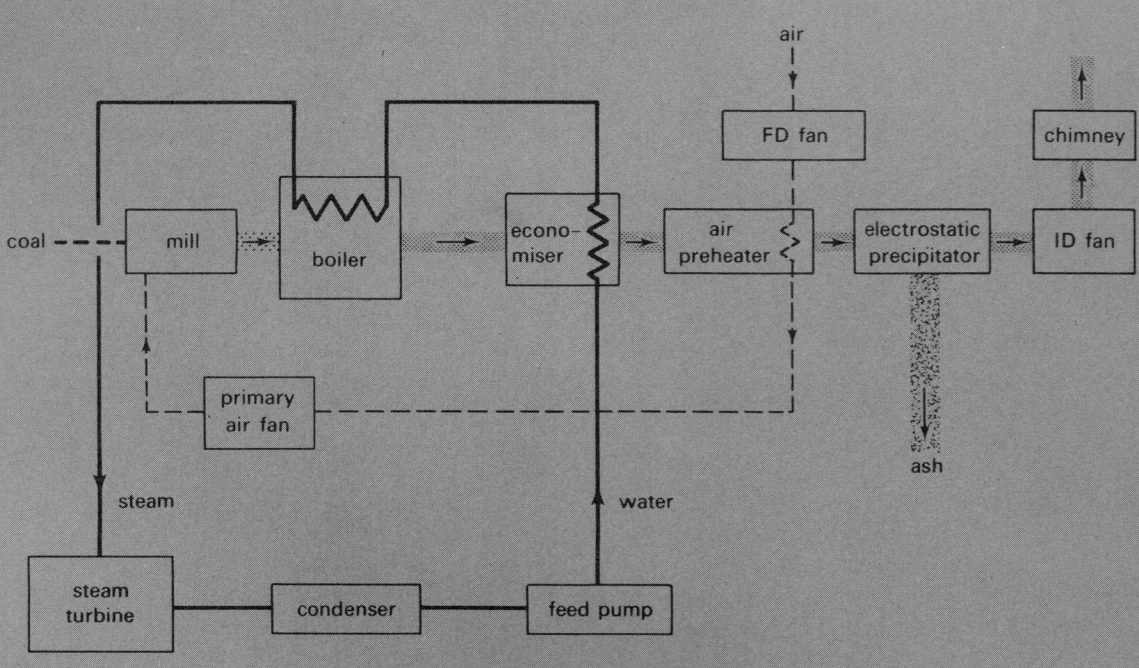

Fig. 4 Flow diagram for a power station

Coal handling

The preparation of the raw coal before it is burned in the boiler furnace is of prime importance amongst the power station services. A modern coal-fired station burns coal that has first been reduced to an extremely fine powder of the consistency of flour, which is known as pulverised fuel or PF. It is capable of being sprayed like a liquid into the combustion chamber as a precise mixture of coal and air at the right temperature and with the correct moisture content. One type of pulverising mill in general use consists of a heavy cast iron turntable over which are suspended two or three heavy special iron-alloy rollers each loosely mounted on an axle (see Fig. 3). The whole of this assembly is enclosed in a massive cylindrical cast iron housing mounted on a substantial base containing the turntable driving gear. Raw coal is introduced into the mill chamber through a regulating feeder at the top of the casing and drops on to the rotating table. The coal moves into the path of the heavy rollers which are rotated by contact with the bed of coal on the table. The arms carrying the rollers are spring-loaded and can pivot vertically, making it possible for the rollers to rise as they make contact with the pieces of coal. Grinding is thus achieved by a combination of impact, abrasion and crushing. The roller is prevented from making contact with the grinding table and the gap between them is regularly checked and adjusted to cater for the wear that takes place. The ground coal is continuously removed from the grinding area by an induced air flow entering the mill through a specially shaped duct; the air passes through an annular space around the perimeter of the grinding table and sweeps upwards through the body of the mill carrying the ground coal to a centrifugal classifier which rejects over-size particles. The larger particles are returned by gravity for further grinding, whilst the coal of required fineness is carried away by an exhauster fan to the burners. A mill of the size shown in Fig. 3 can process 2×10^4 kg of coal an hour.

Pulverised fuel can be rather abrasive and certain parts of the PF system are subject to extremely rapid wear. The grinding table and rollers, which are usually produced from a hard type of alloyed cast iron or cast steel, can wear rapidly, and it is possible for up to 30 grams

The force of friction

'Friction' is the term used to describe the resistance exerted to motion when one solid body is moved upon another. In the case of liquids and gases, friction occurs within the body itself.

Suppose a small horizontal force P is applied to a block of weight mg resting on a rough horizontal surface (Fig. 5a). Four forces now act on the block:

- the weight mg of the block acting vertically downwards
- the horizontal pull of P
- the upward reaction R of the surface (what is the source of this reaction?)
- the horizontal friction force F due to the nature of the surface in contact.

Since the block is in equilibrium, $mg = R$, and $F = P$. If the pull P is increased, F will increase so that F and P are still equal, but although P can be increased indefinitely F will reach a maximum value. If P exceeds this maximum value of F the block will move in the direction of P. The maximum value of F, just before movement starts, is known as the *limiting static frictional force* for the particular pair of surfaces in contact. The frictional force *always* opposes the force tending to move one surface relative to the other.

A similar procedure would reveal that whichever way the block was orientated the frictional force would remain the same (see Fig. 5b).

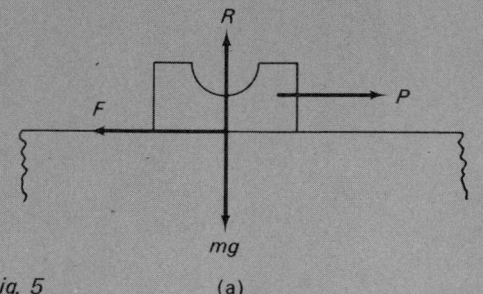
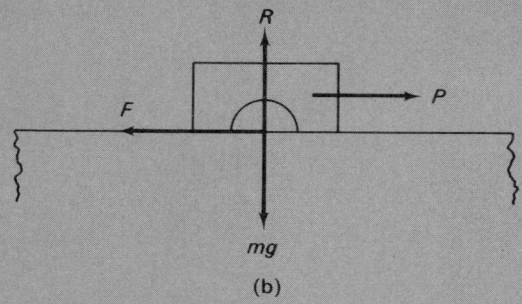

Fig. 5 (a) (b)

of metal to be removed per operational hour on a large mill of the type illustrated. The expected life of these parts is about 8 000 hours. The grinding table, up to 1.8 metres in diameter, and the grinding roller, 1.2 metres in diameter, will cost several hundred pounds to replace. Taking into consideration that as many as thirty mills can be in operation to supply a group of boilers, renewal of such items can significantly affect the cost of maintaining the plant.

The foregoing description applies to a suction-type of milling system i.e. where an exhauster fan induces the flow of air through the mill.

However, the passage of pulverised fuel through the exhauster fan causes wear to take place (see Fig. 7). Current practice, on the larger sizes of mills, is to locate the fan before the mill so that the fan handles only clean air. With this arrangement, the mill operates under pressure, and design changes become necessary in order to prevent the entry of PF into the internal bearing assemblies, and also to ensure that the mill casing is airtight.

The coal mills are normally arranged in groups to serve a particular boiler with fuel, the group being interconnected by a complex system of ducting so that a stand-by mill may be used at any given time. The ducts also suffer from the abrasive effects of the PF; the sections prone to the highest wear-effects are in positions where the direction of flow changes, for example, at the back of a bend in the duct. Such sections are frequently repaired by patching, or are completely renewed if further repair work is not possible. This type of maintenance work becomes almost continuous. Different materials and surface coatings are tried in these locations with varying degrees of success in an attempt to control wear rates. Generally, the maintenance engineer, in collaboration with the research and purchasing departments, has to choose between using expensive materials giving longer service, or relatively cheaper materials which require frequent renewal.

This illustrates an important frictional effect, that the frictional force is independent of the area of contact between the moving bodies. This is known as the first law of friction. The second law is that the frictional force is proportional to the load between the surfaces. If the load mg is varied, the value of the maximum frictional force would be found to be proportional to mg.

The first law may appear paradoxical, but a theory has been put forward which explains the effect of area on frictional force. To start with we must look closely at the nature of the surfaces in contact. All solid surfaces possess a certain degree of roughness: even the surface of a very highly polished piece of steel will be seen under a microscope to have an irregular surface taking the form of peaks and valleys which vary in depth and spacing. Now consider what happens when two such surfaces are placed one on top of the other. Contact will occur only at the peaks, or asperities, as shown in Fig. 6a, and the load must be borne by these points of contact. The load is, however, much greater than the peaks can carry, and they yield until a state of equilibrium is attained, when the areas of contact are sufficient to support the load (Fig. 6b).

Figure 6c shows a single area of contact

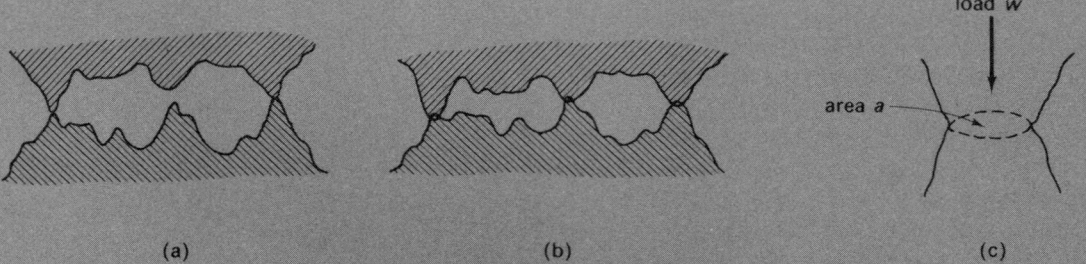

Fig. 6 Deformation of asperities

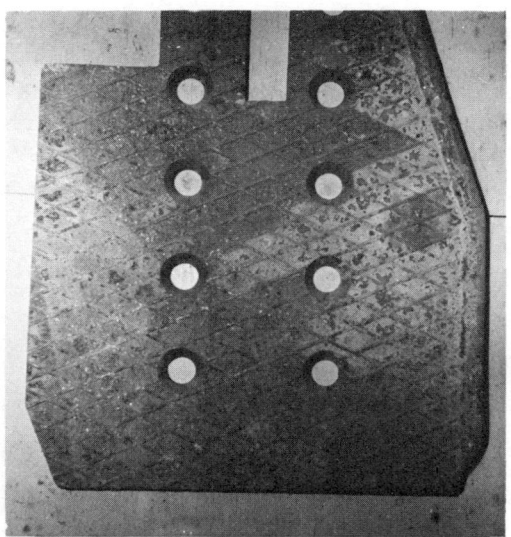 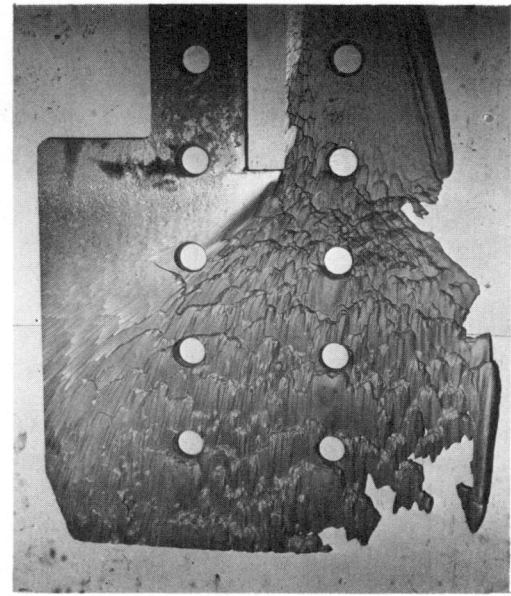

Fig. 7 (left) shows an exhauster fan blade made of 12.5 mm thick cyanide-hardened mild steel. The same blade is shown after 800 hours use (right). Note the wear resulting from the very abrasive nature of pulverised coal. Tests have shown that various forms of surface-hardened mild steel, made by diffusion with chromium, carbon or cyanide, offer no significant improvement on mild steel plate, but the life of the blades can be more than doubled by coating the surfaces with special wear-resistant materials. This is done by welding, sweating or fusing a hard facing on to mild steel sheet. Present exhauster blades incorporates these special coatings.

between two opposing asperities. The direct stress at the contact area is defined as the load w divided by the area a over which the load acts, and in this case is given by

$$\text{stress} = \frac{w}{a}.$$

The stress will be the yield stress (σ_y) in the 'weaker' material because it is the weaker material that has deformed plastically. σ_y is given by

$$\sigma_y = \frac{W}{A} \quad \ldots \ldots \ldots \ldots (1)$$

where W is the total load between the surfaces and A is the final cross-sectional area of all the deformed peaks. Note that A is the area of real contact and not the apparent area of contact.

At the asperities that are in initial contact, the extremely high pressures and rapid deformation cause the materials to cold weld at the areas of contact. The force required to slide one surface over another must be great enough to shear these welded junctions. If this happens, fresh peaks are brought into contact, and these in turn weld together. These welded junctions are themselves sheared, new contacts are made and the cycle is continuously repeated. Every time a junction forms there is a rise in friction, and every time a junction is sheared, the friction drops. This explanation of the friction process is sometimes called the 'stick-slip' theory of friction.

The friction force, then, is the force required to shear all the welded junctions. The shear stress (τ) needed to shear the junctions is related to the friction force F by

$$\text{shear stress} = \frac{\text{shear force}}{\text{area over which shear force acts}}$$

$$\tau = F/A$$

where A is the final cross-sectional area of the junctions or the area of real contact. Hence the frictional force (necessary to produce sliding) is given by

$$F = A\tau \quad \ldots \ldots \ldots \ldots (2)$$

Substituting the value of A from (1) gives

$$F = \frac{W\tau}{\sigma_y} \quad \ldots \ldots \ldots \ldots (3)$$

or $\quad F \propto W.$

Ash handling

The ash resulting from the combustion of pulverised fuel presents almost as great a handling problem as the pulverised fuel itself. Although a large proportion of the stone and pyrites has been removed from the coal during the milling process, there is a residue of some twenty per cent of the original bulk to be disposed of after combustion.

The coarse ash falls to a hopper at the base of each furnace and is sluiced away at intervals to a common sump where it is filtered out. The ash is removed from the sump by a mechanical grab and deposited in a drain area for ultimate disposal. The flue gases, carrying the fine ash, known as pulverised fuel ash or PFA, are drawn from the combustion chamber by induced draught fans. Before being discharged to the chimney, these gases pass through a mechanical separator and an electrostatic precipitator which extract 99.3 per cent of the ash. The pulverised fuel ash is removed from the storage bunker, mixed with water and the resulting slurry is pumped from a collecting pit through a system of pipework to an ash lagoon or to a site for use in a land reclamation scheme. The pulverised fuel ash is highly abrasive and is mixed with water in the final stages to facilitate handling and to reduce wear on the slurry pumps and pipe-work. These pumps, which may be handling over 15 000 litres per minute, will have special abrasion- and corrosion-resistant linings. The sluiceways for conveying the slurry are often lined with cast basalt tiles; cast iron pipework can be used if the slurry has to be transported a considerable distance.

Pulverised fuel firing enables low grades of coal to be used with reasonable efficiency, since the high ash-content is no longer the disadvantage it was a few years ago. The 10^7 kg of coal consumed on average in a 1 000 megawatt power station every day can produce 2×10^6 kg of pulverised fuel ash. The problem of disposing of some 10^{10} kg of pulverised fuel ash annually is now considerably eased by the use of the ash as a construction material of great versatility. The year 1970—71

It is important to realise that the actual area of contact between surfaces is very much smaller than the apparent area, and adjusts itself according to the load. The area of real contact is independent of the way the body is orientated. The area of real contact between two steel plates may be as little as 1/100 000 of the apparent area!

Over a wide range of values of the load W between two surfaces, F is proportional to W, but for low and high values of load the variation between F and W changes as shown in Fig. 8. With very high loads the surface may be grossly deformed and normal sliding cannot take place.

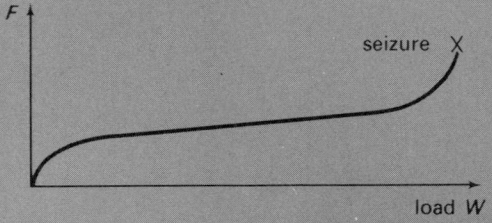

Fig. 8 Variation of friction force with load

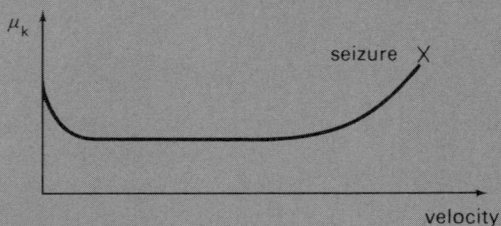

Fig. 9 Variation of kinetic friction with sliding velocity

saw a remarkable increase in the sale of pulverised fuel ash, with over 6×10^9 kg sold for a variety of purposes. It is now widely used in construction work for concrete and mortar mixes, for moulded building blocks and for the production of lightweight aggregate. Large quantities of the ash are used as a bulk filling in the construction of new motorways. The utilisation of a waste product to such commercial advantage is in itself remarkable.

Sometimes the problems that arise from the wear of materials associated with the transfer of abrasive substances, such as coal dust and boiler ash, can be partially solved by using a base material covered with an elastic surface. Pipes and ducts may be lined with rubber. The rubber deforms on impact and absorbs some of the energy of the moving particles, resulting in only a small loss from the rubber surface. Rubber lining, however, eventually weakens and becomes detached from the metal backing; repair is expensive and must be done by specialists. PVC (polyvinylchloride) is frequently used instead but although it has high resistance to chemical attack it has much the same failings as rubber.

Planned maintenance

We have seen that a power station consists of plant employed to move the three basic substances, coal, ash and water, from one part of the station to another. Friction and abrasion will arise as soon as the process of conveying these substances commences, and a power station will therefore begin to wear out as soon as it is commissioned. The pipes, conveyors, ducts and machines will begin to deteriorate from the first few seconds of operation.

In modern industry there is a constant need for new machines and industrial plant to do new jobs in all types of conditions. In a Government publication (*Lubrication [Tribology] Education and Research 1966*) it was estimated that if the basic principles of tribology were carefully applied throughout industry the annual financial savings would be about £500 m. The estimated savings are summarised in Table 2.

Coefficient of friction

As a direct consequence of the second law of friction, we can see that except for very high and very low values of load the ratio F/W for two surfaces in contact is constant. F is the maximum friction force or the force required to start one body moving on the other. This ratio is termed the *coefficient of static friction* and is given the symbol μ.

Hence $\quad \mu = \dfrac{F}{W} \quad$ and $\quad F = \mu W$.

Referring to equation 3 on page 7,

$$F = \frac{\tau}{\sigma_y} W = \mu W$$

hence, $\quad \mu = \dfrac{\tau}{\sigma_y} \quad \dotfill \quad (4)$

Table 1 gives some typical values of μ for various substances.

Once sliding starts, the force required to maintain uniform sliding motion is not so great as that needed to cause sliding in the first place, although this lesser force is again proportional to the normal reaction.

TABLE 1

Values of the coefficient of static friction, μ, for various surfaces in contact. (Reference should also be made to *Structures,* Table 1, page 21).

Materials	μ
Hard steel on steel: dry in air	0.6
lubricated with mineral oil	0.14–0.2
lubricated with organic oil	0.08–0.1
lubricated with molybdenum disulphide	0.1
White metal on steel unlubricated	0.7
lubricated with mineral oil	0.1
Phosphor bronze on steel unlubricated	0.35
lubricated with mineral oil	0.15–0.2
PTFE on steel	0.04–0.1
Perspex on perspex	0.8

The ratio of this force to the normal reaction is termed the *coefficient of sliding friction*, or the coefficient of kinetic friction. It remains constant over a wide range of velocities, but the relationship does not hold for very small or very high velocities (see Fig. 9).

TABLE 2

Estimated savings	£ million p.a.
in energy consumption through reduced friction	28
in manpower	10
in lubricant costs	10
in maintenance and replacement costs	230
in losses consequent upon breakdown	115
through greater plant usage and better mechanical efficiency	22
through increased life of machinery	100
	515

It is interesting to note that during 1971 the electricity supply industry spent in the order of £80 m on the repair and maintenance of power station plant of all types.

How significant is the problem of wear to the maintenance engineer? The rate at which wear takes place will influence the time at which a part of the plant will be taken out of service, and for how long. The engineer must be able to predict when plant should be serviced and lay down a maintenance programme. He will be able to estimate the cost of materials, the amount of labour required, and the loss of revenue resulting from the plant being out of service.

Early in this chapter we mentioned the need for employing the principles of tribology in design, for good design will only result from a sound knowledge of basic principles. Nearly all machines incorporate built-in wear processes because there are always some parts in sliding or rubbing contact. The good designer will exploit his knowledge to the full. He will ensure that he has chosen the correct materials, used the best sort of bearing for the job, seen that the environmental effects will be adequately coped with and he will have given thought to the ease and cost of maintenance.

It is important to realise that the laws of friction are only approximately valid, because of the contamination of surfaces by a film of oxide and absorbed gases. Also this film may have a marked effect on the coefficient of friction. If metal surfaces in contact are chemically clean, the welded junctions are very strong and the friction force can be exceedingly high: tests of chemically clean surfaces have implied values for μ of nearly 100! Because of surface contamination, equation 4 may not hold true in practice.

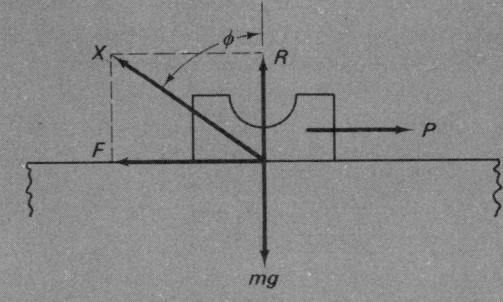

Fig. 10 Angle of friction

Angle of Friction

We have seen that there is an upwards reaction R on a body of weight mg. The resultant X between this reaction and the friction force F needed to move the block is shown in Fig. 10. The angle ϕ between R and X is the angle of friction. From the figure,

$$\tan \phi = \frac{F}{mg} = \frac{F}{R} = \mu \quad \ldots\ldots\ldots\ldots (5)$$

Thus the tangent of the angle of friction is equal to the coefficient of friction.

2 Bearings

The bearing is the fundamental element in mechanics; even the simplest machine, the lever, must have a bearing, or fulcrum. All movements in mechanisms and machines require some sort of bearing to locate and guide the moving parts. A machine designer will try to convert as much as possible of the energy put into a machine into useful output work and will therefore attempt to reduce the energy losses that arise through friction at the bearings. In this chapter we shall be concerned with the design of bearings in common use and their application in a power station and elsewhere.

We must first know what a bearing does. Its functions are to allow surfaces to move relative to one another with the least loss of energy and to locate and guide the moving surfaces. There are two main groups of bearings: those in which the surfaces are in sliding contact (known as plain bearings) and those in which the surfaces are in rolling contact.

Plain bearings

The most common type of plain bearing is the journal or sleeve bearing in which a shaft, or journal, rotates in a sleeve. So that the shaft can rotate in the sleeve and can also be adequately lubricated, the journal is made slightly smaller in diameter than the sleeve.

Plain bearings can be made to carry either a transverse load, or an axial or thrust load, or a combination of both (see Fig. 11). The

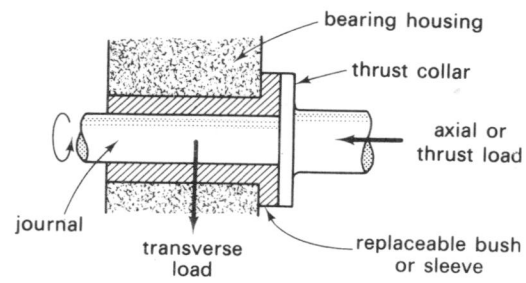

Fig. 11 Simple plain combined journal and thrust bearing

Friction and wear in plain bearings

In the previous chapter, friction was explained by a 'stick-slip' effect at the interaction of two surfaces moving relative to each other. If two metal surfaces are moved, relative to each other, one or both surfaces will gradually be eroded. The process of erosion depends on the nature of the surfaces but, in general, tiny fragments are torn out of the surfaces as the welded junctions shear at the areas of contact.

The wear of the surfaces can theoretically be eliminated by introducing a material weak in shear, known as a lubricant, between them. If the film of lubricant is deep enough the surfaces will be sufficiently far apart to prevent the formation of any welded junctions directly between them. What happens now, if one surface is moved relative to the other? It is obvious that shearing will take place at the weakest part of the interface. The force required to shear a lubricant is very much less than the force needed to shear the welded junctions that would form at unlubricated surfaces and shearing now takes place inside the lubricant film. Also, as the film shears, it is immediately reformed; the process is therefore continuous and there is no 'stick-slip'.

Wear will not be entirely eliminated, in practice; for example when motion is commencing or finishing, the lubricating film will not be sufficiently thick to prevent contact between the solid surfaces.

Bearing material

Let us suppose that an unlubricated bearing sleeve is made of the same metal as the shaft. The welded junctions that are formed are of the same material as both surfaces. As the peaks deform, the surfaces become work-hardened with the result that the shear strength, or resistance to shear, of the welded

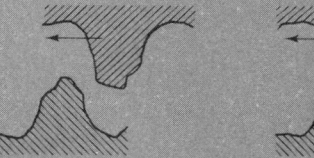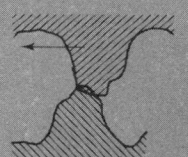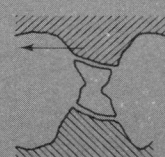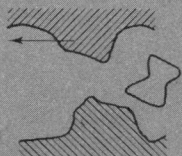

Fig. 12 Interaction of asperities of similar metals in sliding contact

sleeve of a journal bearing is usually held in a cylindrical housing, which may be of cast iron or steel. In order to use a suitable material for the bearing surface the housing may be bushed as shown in Fig. 11. This gives economy of material and the advantage that the bearing may be re-bushed when worn. In many cases the bearing is split, for ease of assembly and repair, and the top portion or cap is secured to the lower section by bolts or studs (see Fig. 13). The bearing bush is also split and removable; these split sections are known as 'brasses' or 'shells'. This does not mean that they are necessarily made of brass; they are usually cast from gunmetal or other alloy, and machined to suit the bearing block.

Brasses are usually lined with white metal, a material which can easily be renewed *in situ* if necessary. White metals are alloys which may contain lead, tin, antimony or copper and are divided into two groups, tin-based and lead-based. The former are known as Babbit metals after their inventor, who first introduced them in 1839. White metals make excellent bearing materials, but because of inherent weaknesses they must be supported by the stronger material of the brasses. The properties of a good bearing material are discussed below in text B.

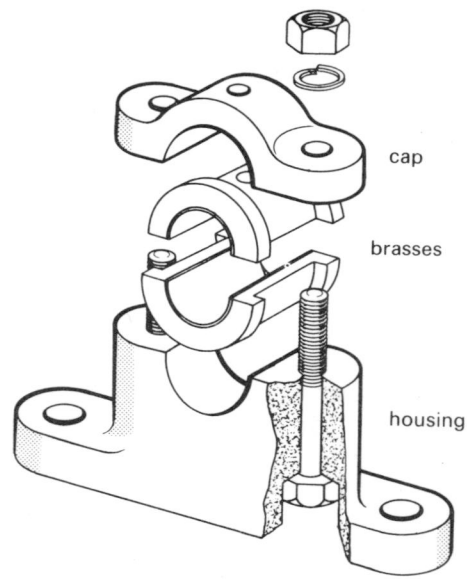

Fig. 13 Exploded view of a common type of bearing

The bushed bearing is easily adjusted for wear. Thin strips of metal known as shims may be inserted between the brasses when first fitted and these are removed to take up the wear; alternatively the white metal can be renewed, bored to almost the correct size, and

junctions, is greater than that of the metal on either side. As it is more difficult to shear these strong welded junctions, shearing will occur in the bulk of the metals, and large particles will be torn out of the surfaces (see Fig. 12). For this reason engineers are reluctant to allow sliding contact between similar metals, and bearings are therefore always made in a different metal from that of the shaft.

Suppose the bearing metal is softer than the metal of the journal. Experiments have shown that in the case of a soft tin-based alloy sliding on steel, shearing will take place at the welded junctions that form at the points of contact between the surfaces. This is because the junctions are weaker than the materials on either side of them. There is very little wear as there is very little transfer of metal from the surfaces.

If the bearing material is a lead-based alloy the welded junctions are stronger than the soft alloy and the bearing wears because shearing takes place a little distance within the softer material, with the result that tiny fragments are torn out of the surface of the alloy. The small loss of material due to wear is from the bearing face and not from the journal. Wear at the bearing is far easier and cheaper to rectify than wear of the journal; also the softer material is able to embed some of the fragments, preventing scoring of the journal.

It is chiefly for these reasons that bearing surfaces are usually made of soft alloys such as white metal; but there are other advantages, for white metals have many of the properties required of a good bearing material. The properties which make white metals suitable for bearings, apart from those mentioned above are:

they conform to the shape of the shaft, and scratches become smoothed over
they easily embed small abrasive particles which may find their way into the bearing
because of their tin or lead base they have high resistance to corrosion
they have low melting points (180°C –250°C depending on their alloy constituents), so that in the event of a lack of oil or grease reaching the bearing the white metal will melt, making a larger clearance between the hot shaft and the bearing, thereby preventing seizure.

White metals, however, are weak and cannot support heavy loads; they also have low fatigue strength. For these reasons white metals in bearings are usually backed by stronger metals held in a rigid housing. (Fatigue is discussed in Chapter Seven of *Structures*).

Improvements in powder metallurgy have led to

scraped to fit the shaft. On small bearings the entire brasses are usually replaced.

Turbine generator main bearing

The rotor of a steam turbine is supported by two or more main journal bearings of the white-metalled type. These are lubricated by an oil-circulating system (see Chapter 3). Each bearing comprises a bronze or cast iron shell made in two halves bolted together and lined with white metal. Often four removable steel pads, spherical on the outside, are fixed to the shell (see Fig. 14). The bearing shell rests on the pads in a spherical seating in the turbine bed in order to permit alignment with the shaft. Shims can be added under the pads to ensure that the rotor shaft is in the correct position.

Thrust bearings

It often happens that in the design of a machine in which the principal load is a transverse one, some provision has to be made for a considerable axial load. This occurs, for example, in the steam turbine, and a thrust bearing, or combined journal and thrust bearing, is fitted to resist the axial load. In turbines of the double flow type the total thrust on the rotor is reduced by allowing the steam to expand along the turbine in opposite directions.

A thrust bearing is essentially a bearing surface at right angles to the axis of rotation and in the case of the steam turbine resists the thrust that arises from the reaction of the rotor blades to the pressure of the steam. Also the thrust bearing fixes the axial position of the rotor relative to the cylinder so that there are

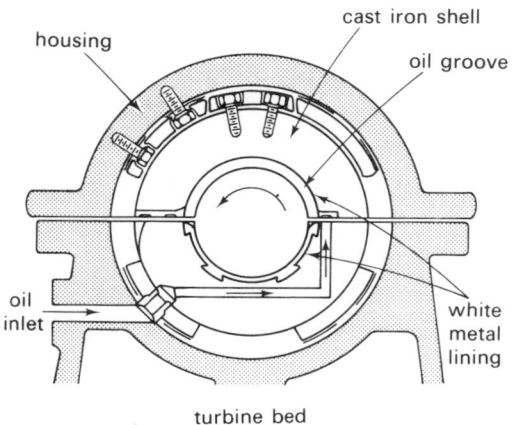

Fig. 14 Section through a turbine main bearing

the widespread use of porous bearings. Porous bearings are made by 'sintering' compressed powder, usually bronze or iron. Powdered metal is compacted and heated in a reducing atmosphere to prevent oxide formation: the process causes the powder particles to weld together. The porous metal structure produced can absorb from 15 to 35 per cent by volume of oil which lubricates the bearings for very long periods before replenishment. Porous bearings are available cheaply and in a range of standard sizes. Their main disadvantage lies in the reduced mechanical properties caused by the porosity; they are not, therefore, suitable for heavily loaded bearings.

In some cases a low friction material such as polytetrafluorethylene (PTFE) is used as a bearing material. These bearings are usually run dry, cutting out maintenance and lubrication altogether. PTFE is, however, mechanically weak and has a high coefficient of expansion; it would seize on a shaft if it were allowed to run hot. It is sometimes incorporated in a sintered porous bearing, giving the advantages of both the surface friction properties of PTFE and the strength and heat conductivity of the sintered metal.

The tilting pad thrust bearing

Reference to Fig. 15 will show that a large turbine thrust bearing comprises a number of tilting steel or bronze pads. This type of bearing is known as the Michell thrust bearing, after its inventor. The principle of its operation relies on hydrodynamic lubrication which is dealt with in more detail in Chapter Four. The bearing is filled with oil and as the shaft rotates, the pads tilt and a wedge-shaped film of oil forms between the face of each pad and the thrust collar. Sufficient pressure is produced by the converging oil film to balance the thrust load. The pivoted pads take up a position dependent upon the speed of the shaft, the axial load, the dimensions of the bearing, and the viscosity of the oil. The taper of the oil wedge is always very small, usually about 1 in 2 000 to 1 in 1 000. The mean thickness of the oil film between the pads and collar is approximately 0.05 mm—0.1 mm. Pads are also provided on the opposite side of the bearing to provide for any accidental reversal of thrust.

Rolling friction

Pure rolling action in which, say, a cylinder rolling on a plane surface makes contact along a single line, never occurs (see Fig. 16a). All materials are elastic to some degree and deform under a load, so that there is always an area of contact between a cylinder and a

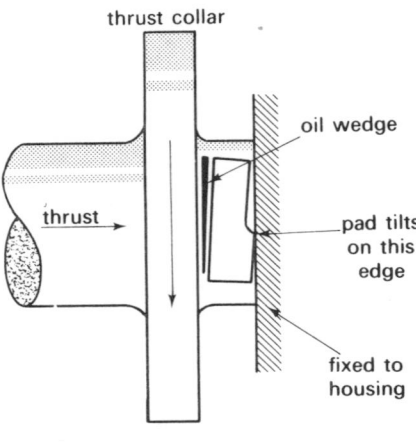
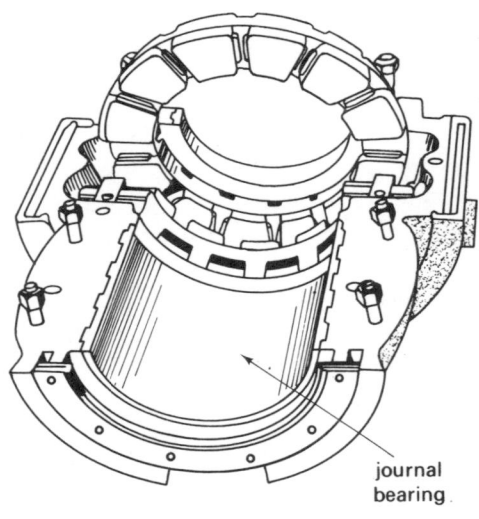

Fig. 15 (a) Operation of thrust bearing   (b) Combined journal and thrust bearing

sufficient clearances between the moving and stationary parts. These clearances may be as small as 0.25 mm. Any change in the axial position of the rotor with respect to the turbine may wreck the complete turbine assembly. Figure 15b shows a combined journal and thrust bearing for a small turbine.

Roller bearings

It is common knowledge that it is easier to roll rather than to slide surfaces over one another. The coefficient of friction of a steel wheel rolling on a steel rail is as low as 0.002. For example one man can keep a 100 kN

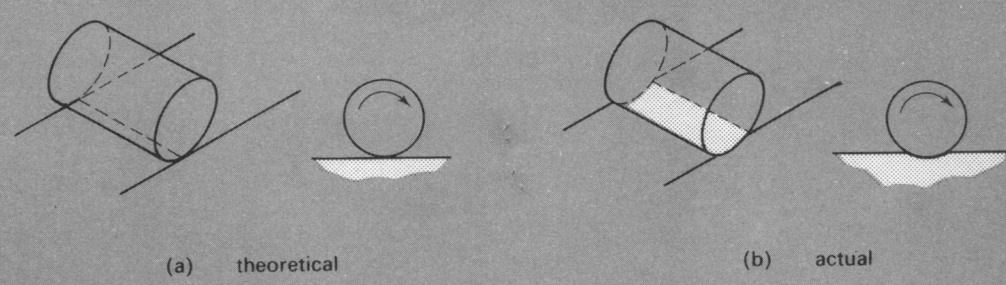

(a)   theoretical   (b)   actual

Fig. 16 Rolling a cylinder on a plane surface

plane surface. This effect is shown highly exaggerated in Fig. 16b.

Experiments have shown that any slip which may occur between the roller and the surface is inadequate to account for the frictional losses. How, therefore, does rolling friction arise, if it is not explained by sliding? When the cylinder is pressed into the plane surface the surface deforms; the cylinder also deforms but we will assume its deformation is small compared with that of the surface. As the material is elastic the surface returns to its original shape when the cylinder is removed.

There is, however, a loss of energy owing to what is known as 'hysteresis'. Most of the mechanical energy absorbed when the surface is deformed reappears when the cylinder is removed, because the material is elastic. However, a small proportion of this energy is converted into heat due to interaction of the atoms as the deformation processes take place. This heat is dissipated in the bulk of the material and the energy loss is known as 'elastic hysteresis' loss. It can be thought of as being a sort of internal friction effect in the material, and it is this loss of energy when the materials deform that accounts for the rolling friction. This explains why rolling friction is scarcely affected by lubricants. (Why, then, are lubricants always used with bearings?)

Types of rolling bearings

The most common rolling bearing is the single row deep groove ball bearing which is suitable for transverse

wagon-load of coal in motion on a level track, a feat which would be impossible under conditions of sliding friction.

A rolling bearing consists of inner and outer rings, or races, with the rolling elements between them. These rolling elements are invariably of steel and are spherical, cylindrical, barrel-shaped or conical depending on the type of bearing. They are prevented from rubbing against each other by a cage, which also retains them so that they do not fall out when the bearing is handled. The inner ring is attached to the shaft and the outer to the housing casing. The rings are usually a tight or 'interference' fit with the shaft and casing. Ball and roller bearings are not located by set screws or keys, because the severe local stresses which are set up may distort the races.

The design of a rolling bearing rarely incorporates any facility for adjusting to wear. Examples of adjustable ball bearings are the 'cup and cone' types found in bicycles and some motorcycles. Wear in rolling bearings is microscopic until just before the bearing fails; then the wear rate is rapid and the whole unit must be replaced. Provided the bearing is correctly mounted, lubricated, protected and properly handled, ultimate failure is by fatigue of the material. (Fatigue is the name given to the progressive deterioration of the resistance of a material to repeated stresses.) Replacement costs of rolling bearings are high compared with the re-metalling of sleeve bearings, but the provision of adequate lubrication is easier. For slow speed bearings, grease can be used and the initial charge can sometimes last the life of the machine.

Linear bearings

The types of sliding bearings described so far are those most commonly used, but we must discuss another important sliding bearing: the linear bearing.

This type of bearing is to be found on machine shop equipment such as lathes, milling machines and shapers. All these machines have parts which move relative to one another, usually in order to change the position of the cutting tool relative to the workpiece being machined, and it is the slides and slideways of the machine which guide and locate these parts. Figure 18a shows a common form of machine tool slideways.

and thrust loads, but must be accurately aligned with the shaft. A variation of this type is the self-aligning ball bearing which has two rows of balls each running in its own groove on the inner race and sharing a spherically shaped track in the outer race. The inner race and ball set can adjust themselves to the bearing axis and the bearing will therefore cater for misalignment on assembly or deflection of the shaft under load, and (in the long run) may reduce maintenance costs.

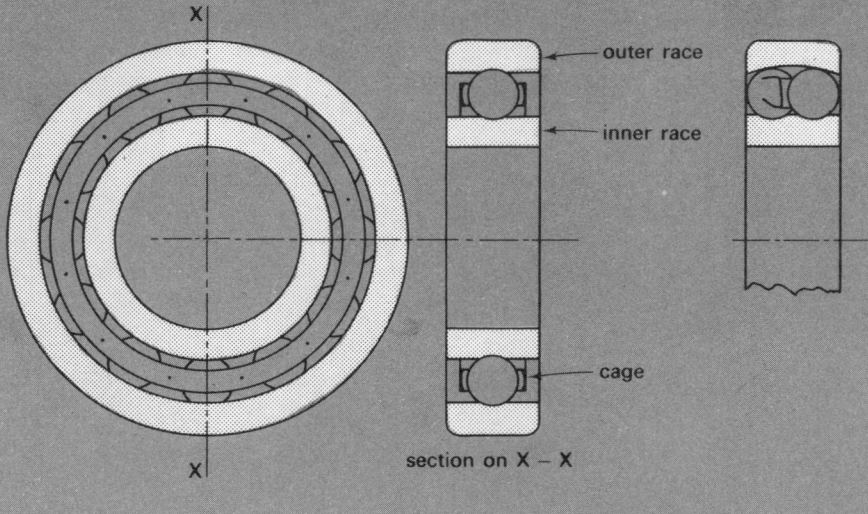

(a) single row deep groove ball bearing

(b) self-aligning ball bearing

Fig. 17

Wear in bearings of this type is extremely critical, as the quality of the work produced is largely dependent on the accuracy of the slideways and if the surfaces become worn beyond a certain point the machine no longer produces accurate work. Sometimes rollers are inserted between the surfaces to make a linear rolling bearing (see Fig. 18b), and the force required to move the slideways is much less than that for a corresponding sliding linear bearing.

It may seem at first sight that slide surfaces should be as flat as possible. But if they were, the surfaces would tend to adhere (like the 'wringing' together of slip gauges) and the surfaces are, therefore, ground or scraped to give areas of very slight concavity.

On modern large machines the slideways are of the hydrostatic type, where oil is fed to

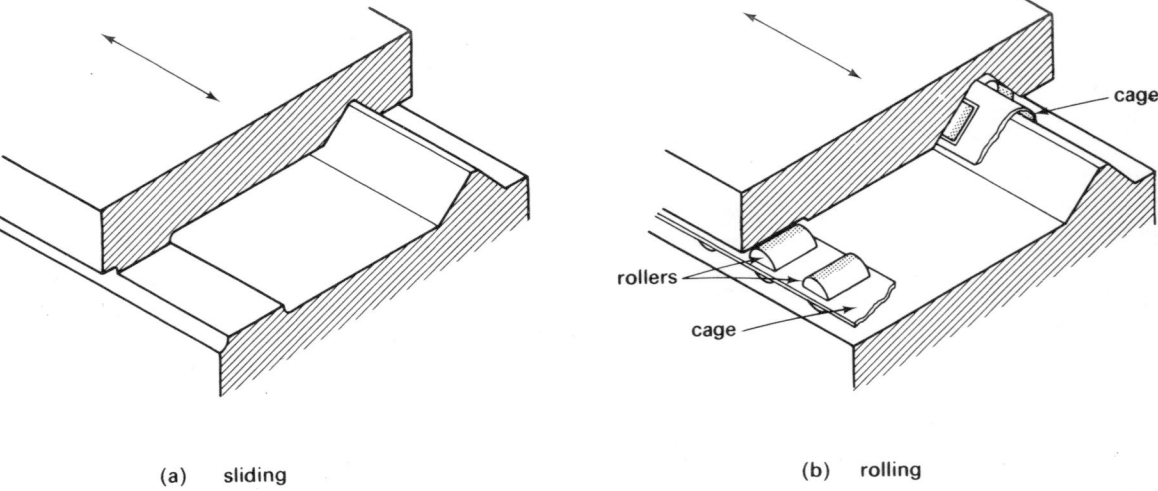

Fig. 18 Types of machine tool slideways

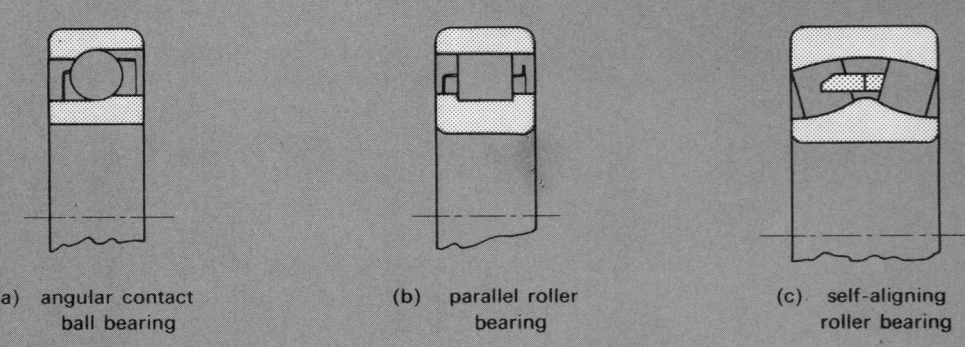

Fig. 19

The angular contact ball bearing is designed so that the line of contact is always at an angle to the radial plane of the bearings. It is therefore able to withstand combined axial and transverse loads. These bearings are usually used in axially opposed pairs, care being taken to ensure they are facing the right direction in order to take the thrust.

The parallel roller bearing is a type suitable for moderately high speeds, ($10^3 - 10^4$) rev min^{-1}, and heavy transverse loads and requires a high degree of accuracy in its manufacture. It is possible for the inner race to move axially relative to the outer race, thereby permitting the bearing to take up expansion due to the effects of temperature. Self-aligning roller bearings incorporate a spherical track on the outer race and barrel-shaped rollers. They can withstand higher transverse loads than the self-aligning ball bearing.

the slideways under pressure. A constant film of lubricant of accurately controlled thickness flows between the moving surfaces (see page 41).

Choice of bearing

The engineering designer will select with care the type of bearing best suited to a particular purpose. He will have to consider:

  the load and stresses it will have to sustain,
  production costs,
  its efficient lubrication,
  the ease and cost of installation and maintenance.

Heavily loaded rotating parts usually require journal-type bearings with separate or combined thrust bearings. Examples are bearings for steam turbines, large centrifugal fans and boiler feed pumps.

For lighter loads rolling contact bearings are often used in combinations of different types. In some cases, e.g. a pulverising fuel mill table drive, combinations of plain and roller bearings are used.

Porous oil-impregnated bearings are often used for light loads and low shaft speeds, as are dry bearings made of PTFE (polytetrafluorethylene) or other low-friction material. Steel-backed PTFE is sometimes used for sliding expansion joints in bridges.

TABLE 3

Coefficients of static friction, μ, for plain and roller bearings

Type of bearing	μ
PTFE (dry)	0.05–0.25
Graphite in metal	0.08–0.2
Copper impregnated with PTFE	0.05
White metal on steel, unlubricated	0.7
White metal on steel, fully lubricated	c. 0.001
Self-aligning	0.0010
Cylindrical roller	0.0011
Thrust ball	0.0013
Single row, deep groove	0.0015
Needle bearing	0.0045
Taper roller	0.0018

A variation of the parallel roller bearing is the needle roller bearing. It is less efficient but can be used where space is limited. This type of bearing is used for low rotational speeds and oscillating (back and forth) motions, such as pivots, universal joints and rocker arms.

Taper roller bearings, as in Fig. 20b, carry both transverse and thrust loads. The tapers of the races and rollers have a common apex on the axis. Where axial loads are high, a bearing with a steep taper angle is chosen, the usual range being from 7½° to 30°. As with the angular contact bearings, taper roller bearings are usually mounted in pairs.

Figure 20c shows a single thrust ball bearing which will take a thrust load only. Greater loads can be carried on the tapered roller thrust bearing, Fig. 20d. In cases where correct alignment is difficult, a self-aligning spherical roller thrust bearing is available.

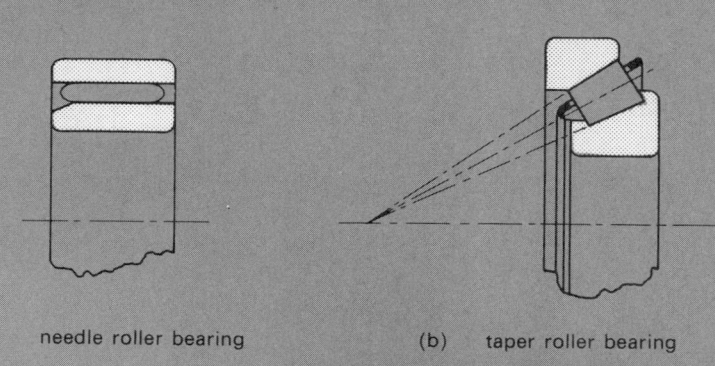

(a) needle roller bearing

(b) taper roller bearing

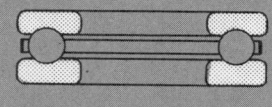

(c) single thrust ball bearing

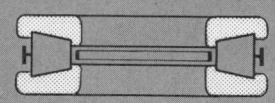

(d) taper roller thrust bearing

Fig. 20

3 Lubrication systems

It was shown at the beginning of Chapter Two that a lubricant interposed between surfaces in moving contact reduces friction and wear. It is vitally important that machines are properly lubricated so as to keep them in good condition and prolong their working life. The various practical methods of applying lubricants to machine parts will now be discussed.

For many years in textile mills and engineering shops the 'oiler and greaser' was a familiar figure, responsible for the routine lubrication of the spinning frames, looms and line shafting in the mills or workshops, doing little else but supply oil and grease to the mill machinery as an almost continuous maintenance cycle. In a modern 2 000 megawatt power station the machinery ranges from a simple power cylinder operating a boiler control damper, which may only require greasing periodically, to a huge 500 megawatt turbine-generator requiring a complex lubricating system distributing oil to many points on the machine. The system must be fully automatic and, of necessity, almost 100 per cent reliable, as cessation of flow for only a very short time could result in serious damage to the machine and loss of generating capacity, perhaps at a critical period. A turbine generator will be expected to be absolutely reliable over a period of many years.

Oil lubricators

The method of lubricating a machine or mechanism will depend upon, among other things, the job it has to do, its operating speed and the size of components. Plain bearings of simple mechanisms such as hinges and pivots, which operate intermittently or at very low speeds, may be lubricated merely by injecting oil through a hole in the bearing cover. This is a primitive method which may introduce dirt with the lubricant and, possibly a more serious drawback, it requires the attention of maintenance staff. Dirt and moisture can be excluded by enclosing the hole in some way

Lubricants

What do we want a lubricant to do? It is intended to prevent wear on the sliding surfaces, reduce the forces necessary to make the surfaces slide, thereby reducing the energy loss at the bearing points, and to cool the sliding surfaces. However, a lubricant must also fulfil a considerable number of other requirements if it is to be suitable in actual service, for it must be stable and effective under all normal operating conditions. What does this mean in terms of the properties of the lubricant, and what sort of substances will be suitable? A lubricant must prevent contact between the actual sliding surfaces, and so must adhere well to the surfaces; this requires that it has a minimum surface energy at the interface (see the section on surface energy on pages 25 to 29). At the same time, the lubricant itself must have low intermolecular forces so that lubricant molecules will slide readily over one another, i.e. during sliding, rupture will occur within the lubricant and not at its interface. This means that the lubricant must have a low viscosity (see Chapter Four).

Surface failure may be due to corrosion as well as to mechanical wear. Thus a lubricant must not react chemically with the surfaces or any other material with which it may come into contact, particularly water and the oxygen in the air. The stability of a lubricant is a much more stringent requirement than appears at first sight. We want the viscosity and surface energies to change as little as possible over the whole working range of temperature and pressure. The physical state of the lubricant must not change: it must not freeze solid, or evaporate appreciably, or entrain gases to form foams (since gas bubbles are non-lubricating and interfere with liquid flow); also the lubricant must retain its properties over a long period of time.

There are still other requirements to be considered. For example, cooling will be more effective if the lubricant has as high a heat capacity as possible. For safety reasons the flash point (i.e. the temperature at which the vapour will ignite spontaneously under 'ideal' conditions) must be beyond the working temperature range. And the cost must be sufficiently low to make the purchase of the required quantities an economic proposition.

This list of desirable properties is so long that there is no single substance which satisfies all requirements completely. The problem is further complicated by the wide variety of situations in which lubricants are required. Consequently, the choice of a lubricant is always a compromise. This compromise is highlighted by the fact that in some circumstances

and the attention required can be greatly reduced by providing a reservoir of lubricant which will feed the bearing surfaces with an adequate supply of oil and will, ideally, indicate when replenishment is necessary. These essentials are met by the wick, or syphon feed lubricator (see Fig. 21), in which oil is conveyed continuously by the capillary and syphonic action of the wick from an oil cup to the point of application; this method has the advantage of filtering the oil supply. The drip-feed lubricator, shown in Fig. 21, supplies drops of oil to the bearing at regular intervals. A needle valve controls the rate of the oil supply and a sight glass provides a visual check on the operation. A lever closes the valve when the oil supply is not required.

In large machines having many bearings or points to be lubricated a gravity-feed oil system may be used. The installation consists of a supply tank, or manifold, which feeds oil directly to the various points to be lubricated, each point having a drip-feed or wick-type lubricator. The tank must be at such a level that all the lubricators receive oil at an adequate working pressure. If this is not possible a pump may be used to pump oil from a reservoir to the bearings. This sort of system is called a force-feed or centralised oil system. Gravity and centralised systems provide dependable trouble-free methods of supplying the correct quantities of lubricant to the bearing surface.

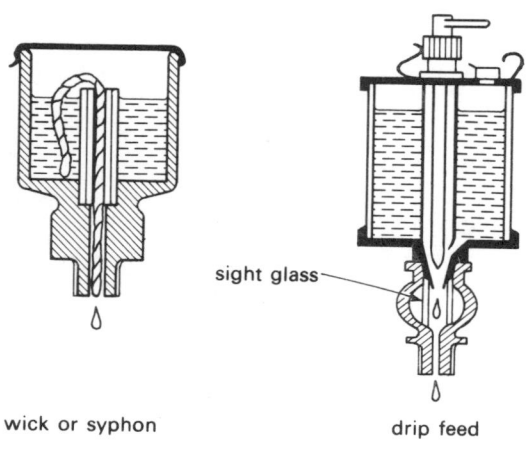

Fig. 21 Oil lubricators

viscosity needs to be high enough to prevent serious leakage of lubricant under pressure, although the viscosity should, as stated above, ideally be low.

Lubricating oils

It is common knowledge that the vast majority of lubricants are 'oils' obtained from crude oil. Simple lubricating oils belong to the chemical group of components known as hydrocarbons. These are compounds of carbon and hydrogen. Carbon is almost unique among the elements in its ability to combine to form long chains, so that with hydrogen it can form a wide variety of compounds which differ in both the number of carbon atoms forming the basic chain, and in the amount of branching there may be in the chain (see Fig. 22). Because of their similar chemical structure, their physical properties are also alike.

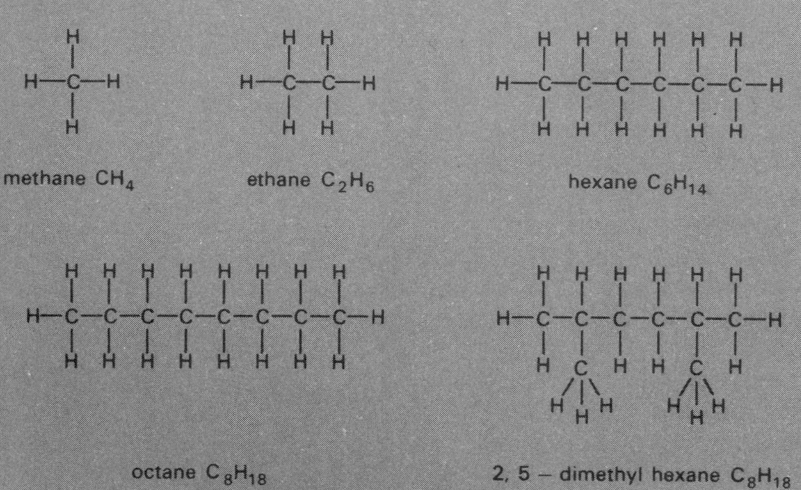

methane CH_4

ethane C_2H_6

hexane C_6H_{14}

octane C_8H_{18}

2, 5 – dimethyl hexane C_8H_{18}

Fig. 22 The structure of some hydrocarbon molecules

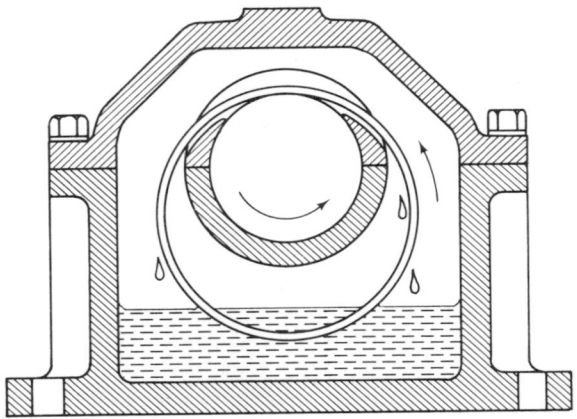

Fig. 23 Ring oiler

For large bearings, such as the bearings on small turbines, induced draught fans and boiler feed pumps, ring oil lubrication is used. This is a simple type of lubricating unit used on journal bearings, consisting of an oil sump in the base of the bearing housing and a ring which encircles the shaft and rests loosely on it when stationary (see Fig. 23). In operation the ring is set in motion by the revolving shaft, and oil is automatically raised from the sump to the top of the journal, where it is dispersed over the length of the journal by channels in the bearing surface. If the bearing is located in a position where the oil in the sump may become overheated, water cooling may have to be provided to maintain the oil at operating temperature. Shafts transmitting up to 25 kilowatts are usually fitted with one oil ring per bearing; larger machines with correspondingly larger bearings may be fitted with two oil rings per bearing. (Can you suggest why oil ring bearings are unsuitable for either very low or very high speeds?)

The lubrication of gears

Mechanical gearing is essential to many engineering machines and systems. It is used to transmit power, to increase or reduce the speed of a drive, or to change the direction of the axis of a drive. A common application lies in matching the speed of an electric motor to a slow-speed and high-torque machine such as the turntable on a coal pulverising mill (see Fig. 3, page 4).

Theoretically, gear teeth mesh along a line

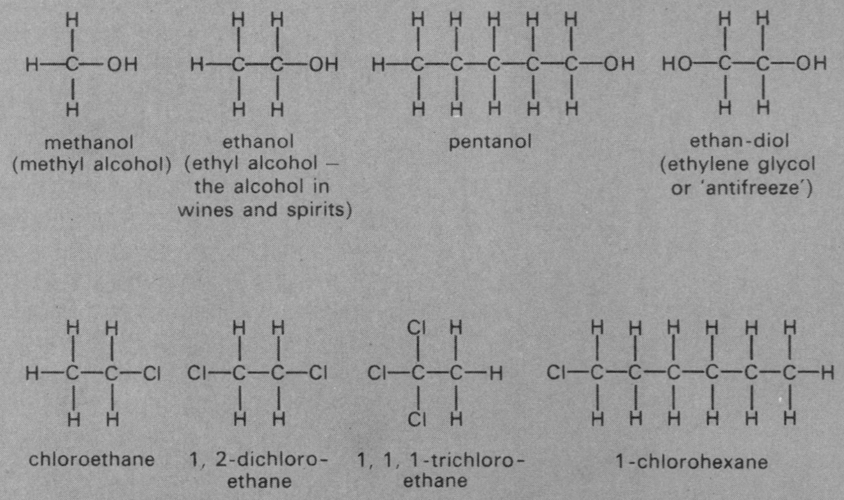

Fig. 24 The structure of some other common carbon compounds

of contact; in practice there is a small area of contact, due to deformation of the gear teeth as they mesh, but as the drive is being transmitted through this small area the pressure on the gear teeth is often extremely high. All gearing is a combination of sliding and rolling friction, and at the high pressures encountered, severe damage can arise if metal-to-metal contact occurs. In addition a high proportion of the energy put into turning the 'driving' gear will not be given to the 'driven' gear. This is because at the area of contact heat energy will be generated due to the slight deformation of the gear teeth (see *elastic hysteresis loss*, in text B on page 14) as they move into contact and also because energy is required to shear the welded junctions that will form during sliding friction. The purpose of the lubricant, then, is to maintain a film that separates the surfaces at all loads and speeds, and which will reduce the energy losses.

For low- and medium-speed gear drives a splash system is normally used (see Fig. 25). The lowest gear dips into an oil reservoir at the bottom of the gear-box casing, and the teeth that have dipped in the oil carry it round to

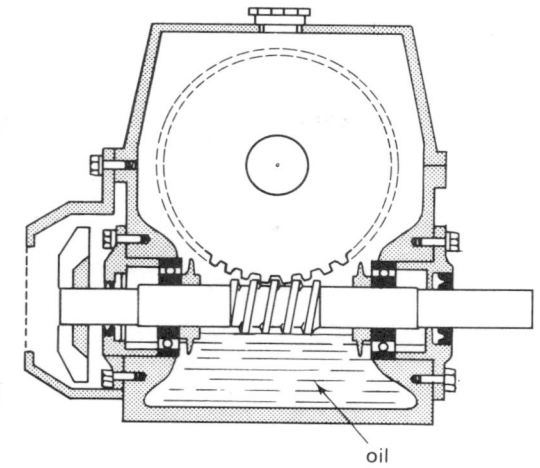

Fig. 25 Splash-lubricated worm gear

the area of mesh. Where the speed is high there is sufficient splash and spray thrown up to be collected in troughs and led to the shaft bearings, or cascaded on to the upper gears. At very high speeds splash lubrication is inefficient; the oil tends to be flung off the gears and churning of the oil may lead to overheating.

Physical properties such as the melting and boiling points and the fluidity of the liquid oil depend on the magnitude of the molecular movements. Thus they are directly related to the mass of the molecule and to the forces between them. Other physical properties depend on the shape of the molecule and on the distribution of charge on its outer surface. These include such properties as surface tension; at the surface, the regular shape, and the absence of atoms with a highly polarised charge distribution such as chlorine and oxygen, enable the molecule to come very close to any solid surface (surface tension is dealt with later in this chapter). The resulting intermolecular force will therefore be large. Chemical properties also depend very largely on the presence of chemically reactive centres or assymetry within the molecule. The hydrocarbons obtained from crude oil, being symmetrical and without reactive centres (compare Fig. 22 with Fig. 24) are called paraffin hydrocarbons. Indeed, the derivation of the word paraffin means 'little affinity', and these hydrocarbons are notable for their low chemical reactivity.

As a consequence we can see that the hydrocarbon oils are a group of compounds from which materials of suitable melting points, boiling points and viscosities can be selected; they are readily and cheaply available from crude oil and have very little chemical reactivity.

Alternative lubricants

What other compounds might have been selected as lubricants? Carbon will also form a wide variety of compounds such as the alcohols and chloroparaffins illustrated in Fig. 24 which contain other elements in addition to hydrogen. The chlorine and oxygen atoms are, however, both much larger and much more reactive than hydrogen, so that compounds containing these elements have a much lower surface tension and a much higher tendency to corrosion and other chemical reactions than do the hydrocarbons.

Water is another possibility that should be examined. It is certainly cheap and widely available, it has the highest thermal capacity of any liquid, a low viscosity, a high surface tension in contact with clean metal surfaces, it is entirely non-combustible, and it has a reasonably high boiling point. In fact it is quite a good lubricant, and in some designs for central heating and many automobile cooling systems no other lubricant is necessary. However, water is unsatisfactory in one major aspect: it corrodes many metals and this corrosion is particularly rapid if the water is contaminated, or if there are several different metals in contact with it, for, unlike hydrocarbons, water readily dissolves most ionic compounds.

Many vegetable and animal oils or fats can be used

High velocity gears are, therefore, usually lubricated by a pressure system where jets of oil are sprayed on to the gears at the engaging or parting teeth. If the jet is directed into the meshing zone, violent churning may take place. The greatest cooling effect is obtained by directing the jets at the disengaging side of the meshing zone and enough oil is carried round to maintain more than adequate lubrication at the point of meshing. The oil in a spray system can be continuously filtered, purified, and cooled in the external oil circulatory system.

Grease lubrication

Grease is normally used where oil is unsuitable, e.g. where oil cannot be retained or where the lubricant is to act as a barrier to exclude dirt. Grease tends to act as its own seal, which is an advantage in machines that must not leak oil, such as machines in the food industry. As the grease remains where it is applied, machines can run for long periods before replenishment is necessary. Grease has the disadvantage, however, that it cannot dissipate heat as effectively as oil does. Greases are generally more convenient than oil in rolling bearings where oil would drain away at standstill causing metal-to-metal contact at the commencement of motion. Rolling bearings should, however, never be overfilled with grease, for the swept volume of the rolling elements and cage displaces some of the grease, and the churning action may give rise to overheating.

The simplest grease lubricator is the Stauffer cup type illustrated in Fig. 28a. It comprises a grease-filled cup which is screwed down on to the base. Routine maintenance of the bearing consists of periodically screwing the cap a half-turn or so as long as the cup is full of grease.

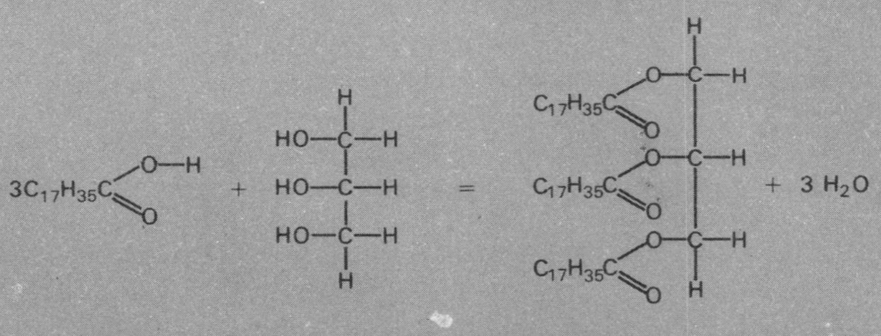

Fig. 26 Spray-lubrication of high-speed gears

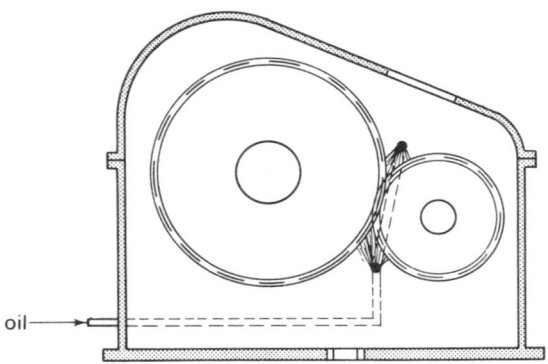

3 molecules of stearic acid + glycerol will form stearin + water

Fig. 27 Stearin is a typical natural fat. $C_{17}H_{35}$ represents a carbon chain as in previous diagrams with 17 carbon atoms and the equivalent number of hydrogen atoms.

as lubricants. These materials usually have a higher viscosity and a greater tendency to chemical breakdown, particularly by oxidation, than hydrocarbon oils. This would be expected from their chemical structure (Fig. 27). Nevertheless their surface tension in contact with metals and low vapour pressure make them superior to hydrocarbons in these respects, and where these properties are important, natural oils may be specified.

A high surface tension becomes particularly important at very low speeds where the lubricant must not drain away, and also where cooling is the prime purpose of the lubricant (as in ball and roller bearings) and good thermal contact is essential. Natural fats are occasionally used in these applications, but hydrocarbon greases are a much more frequent choice nowadays, for they are less prone to oxidation than natural fats, and are usually cheaper, although inferior in some other respects.

Greases consist of hydrocarbon oils with a thickening agent added to give them a semi-solid consistency. This is usually achieved by incorporating materials from the class of substances known as soaps. Most natural oils and fats, and all soaps, are derivatives of a

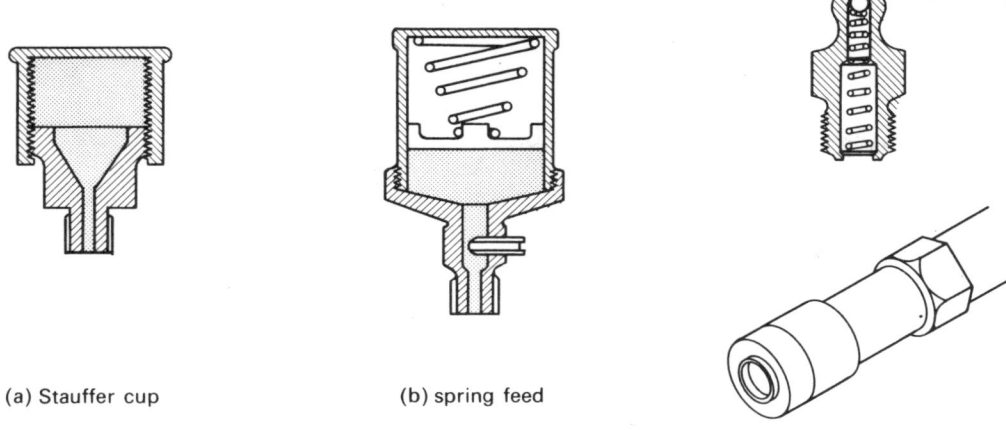

(a) Stauffer cup   (b) spring feed

(c) hydraulic nipple and connector

Fig. 28 Grease lubricators

In cases where lubrication must not depend upon frequent attention, a spring-feed lubricator may be provided as shown in Fig. 28b. Some form of adjuster is provided so that the flow of grease can be regulated.

Alternatively, grease can be periodically injected into the bearing by means of a grease gun in conjunction with grease nipples. These contain a spring-and-ball non return valve to prevent the leakage of grease and the entry of contaminants. Today, however, the manual application of grease is rapidly disappearing, and grease is applied by automatic, centralised force-feed systems which supply the correct amount of grease at regular periods, thus cutting down maintenance and saving time. Also many modern, small, grease-filled bearings are of the 'sealed for life' type and should need no maintenance whatsoever: the complete bearing assembly is renewed after a long period of service.

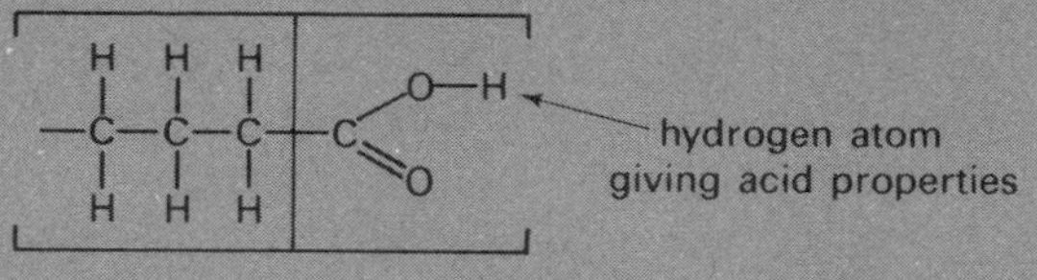

Fig. 29

series of acids of the general structure illustrated in Fig. 29 and in which the chemical bond holding the hydrogen atom responsible for acid properties has been broken and used to join the rest of the molecule to another molecule such as glycerol (see Fig. 27).

If these natural fats are treated with an alkali such as sodium hydroxide, a sodium atom replaces the acid hydrogen; the substance is then called a soap (see Fig. 30). Soaps are quite good lubricants too, as anyone who has unwittingly stepped on a piece can painfully testify. However, their really important property is their strong force of attachment to most solid surfaces. The action of a soap in aiding washing depends upon this as much as upon the more obvious reduction of the surface tension of the surface between water and air, and the same adhesive property of certain water-insoluble soaps is the basis of cosmetic preparations such as lipstick. Soap is used as a lubricant in many metal-forming and pressing operations, owing to this property of surface adhesion.

Sodium-based greases are partly soluble in water, and often make good hand-cleaners, while lithium- and calcium-based greases are insoluble. However, they all have the high surface-adhesion properties which make them useful for lubricant applications.

A realisation of the importance of surface phenomena in lubrication has led to the development of another group of lubricants, which chemically are markedly different from the greases but in which the high adhesion forces and low internal friction within the material itself are still the two important characteristics. Molybdenum disulphide ('Molyslip') is a typical material of this kind. It is a solid in which the layers within the crystal structure are only very weakly bound together, so that internal friction is very low. However, owing to the chemical nature of sulphur, the molybdenum disulphide forms strong chemical bonds with metal surfaces at high temperatures so that an extremely thin but strongly adherent layer of solid lubricant is formed on the metal surfaces.

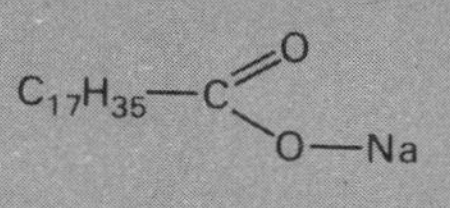

Fig. 30 A typical soap (sodium stearate)

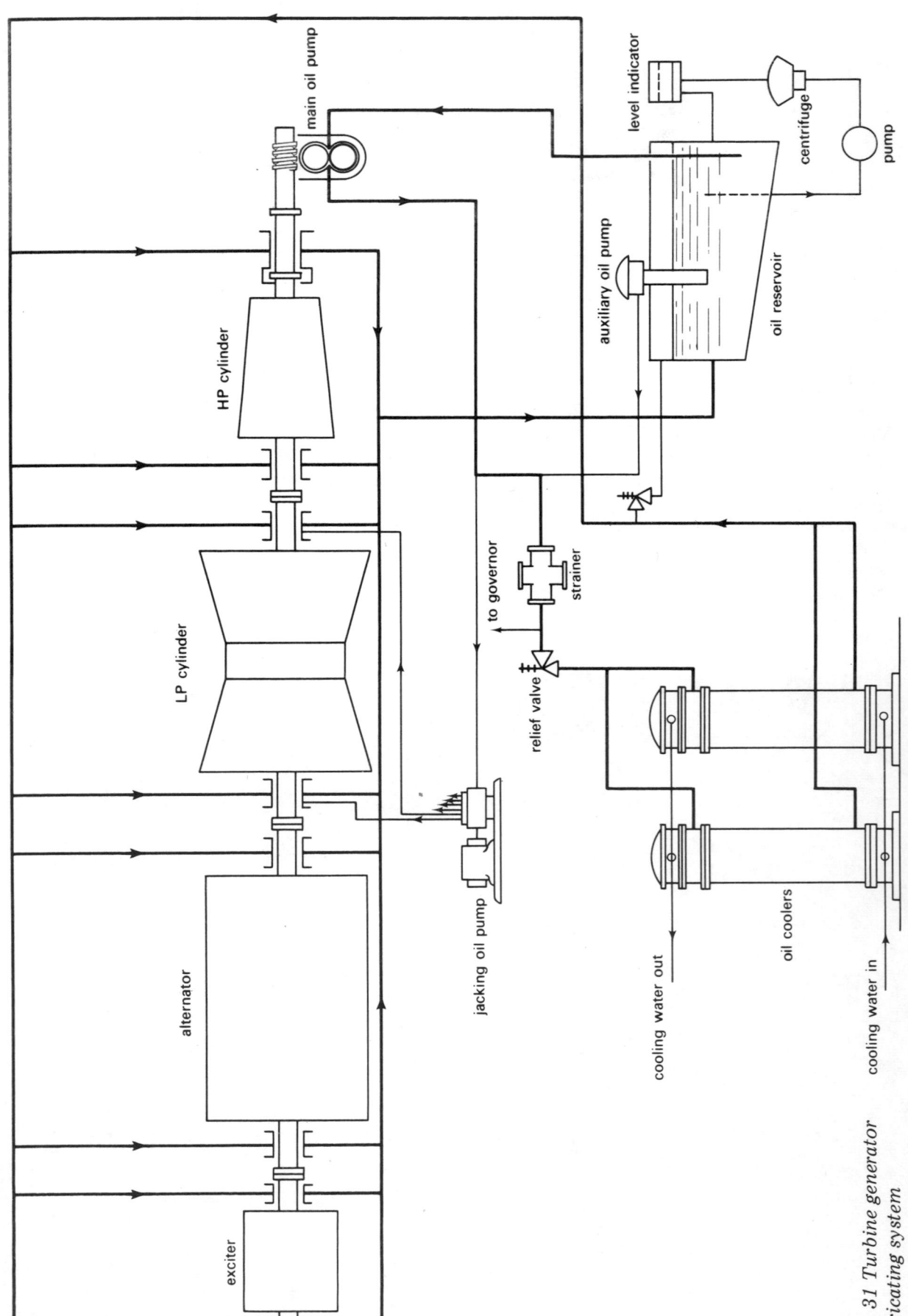

Fig. 31 Turbine generator lubricating system

Turbine generator lubricating system

The details of oil-circulating systems may vary greatly with turbines of different manufacture but all are similar in basic essentials. A typical oil-circulating system for a turbine generator unit is illustrated in Fig. 31.

A main oil pump, driven by the turbine rotor through a reduction gear, draws oil from the reservoir and delivers it under pressure to the main bearings of the rotor shaft. The pressure of the oil at the main bearings is usually within the range of 34 kN m^{-2} to 104 kN m^{-2}. After circulating through the bearing, the oil is returned by gravity to the reservoir. The oilways in the main bearings are designed to allow a greater flow of oil than is required for lubrication only, the additional oil flow serving to remove frictional heat and heat conducted to the bearings from the high-temperature parts of the turbines (steam enters many modern turbines at 568°C). The lubricant thus serves as a coolant to prevent overheating. A temperature indicator is normally provided at each main bearing so that the bearing temperatures can be checked at regular intervals (at least once per hour).

Oil coolers are provided to cool the oil before it reaches the bearings. A typical cooler consists of a cast iron cylinder containing a number of brass or cupro-nickel tubes through which water circulates. Oil flows around the tubes, which are cooled by the water. (The pressure of the oil must be greater than the water pressure: can you suggest why? Also, why are at least two oil coolers always installed?

Before the turbine is started up, a jacking oil pump supplies oil at high pressure (up to 103 kN m^{-2}) to the bearings in order to lift the shaft and ensure that metal-to-metal contact is avoided at the commencement of rotation. An auxiliary oil pump supplies oil to the bearings while the turbine is being run up to speed. A regulator causes this auxiliary pump to cease working when the rotor has reached a speed at which the main pump will deliver sufficient oil to the bearings. The auxiliary is independently driven by a small steam turbine or electric motor and starts automatically when the bearing pressure drops below a pre-determined value; this would happen if the main oil pump failed, or during the shutting-down operation when the rotor is slowing down.

There is another group of solid materials which can act as lubricants too, but they do not depend on adhesion to the lubricated surfaces. These solids include graphite and talc, which have very low internal bonding forces and will therefore shear very readily, and solids such as the waxes and the plastic PTFE, in which the adhesion forces are so low that the welded junctions between a surface and the solid lubricant have very little resistance to shear indeed. These solids lubricate because a relatively thick layer of easily shearable material has been inserted between the two sliding surfaces. Any material which has weak internal bonding will serve as such a lubricant, whatever its chemical composition.

Surface tension

Throughout this book we have been concerned with the frictional forces that arise between two solid surfaces and the forces that occur at the interface between a lubricant and a surface. All these interface forces originate from the same source: the difference between the forces of attraction between molecules within each substance and the forces across the interface separating the two substances. This is illustrated in Fig. 32. The forces may arise in some cases from chemical bonding across the interface but more

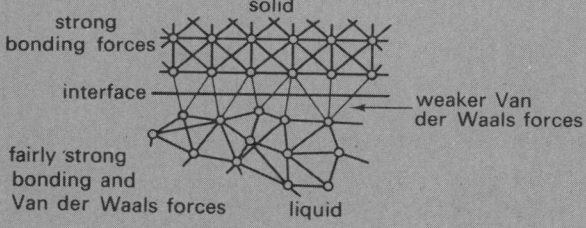

Fig. 32 A solid/liquid interface

usually the interface forces are of the Van der Waals type. It is important to realise that if there are no interfacial forces then there will be no physical surface. Interfacial forces between solids are described as frictional forces, and have already been discussed. Interfacial forces involving liquids are known as *surface tension* forces and it is the purpose of this section to discuss these a little further.

It is extremely important that the oil in the system is kept free from impurities, and purification can be either by filtration or by centrifuge, or both. The centrifugal purifier or 'centrifuge' depends on the principle that if the oil is spun at a high speed the heavier impurities (for example, water) tend to travel outwards. Separation occurs and the outer layer of impurities can be drawn off. The lighter, purer oil is returned to the reservoir. Coarse metallic particles are removed from the oil by the fine mesh incorporated in the strainer, or filter.

The continuous operation of the lubricating system of a turbine generator is so vital that the oil in the system is commonly used to operate the turbine control gear hydraulically. This ensures that in the event of a failure of the lubricating system the machine stops and damage is minimal. A seized bearing on a turbine rotating at 3 000 rev min^{-1} has been known to disintegrate the machine.

Bearing seals

Some method must be found to retain the lubricant at the bearing surfaces. This is normally achieved by fitting some form of seal where the shaft or moving part passes through the bearing enclosure. The type of seal selected depends on the type of bearing and the kind of lubricant used. The seal will also keep out abrasive dust or other substances likely to damage the bearing surfaces or impair the performance of the lubricant; this is important in the case of equipment in wet or dirty locations, such as the coal-conveying and ash-handling plant in power stations. Two methods of sealing are commonly used: the non-contact seal, in which a pre-determined clearance exists between the stationary and moving surfaces, and the contact, or rubbing seal.

Non-contact seals are of three main types: labyrinth; annular groove; and slinger or spinner.

The labyrinth seal consists of a stationary concentrically grooved flange opposite a similar flange rotating with the shaft, the ridges of one flange coinciding with the grooves in the other. The combined effect of this is to present a path of high resistance to the flow of lubricant in one direction and to the entry of dirt and moisture in the other. The labyrinth seal is generally used for grease lubrication but can be designed for use as an oil seal, in which case the clearance between the seal components is

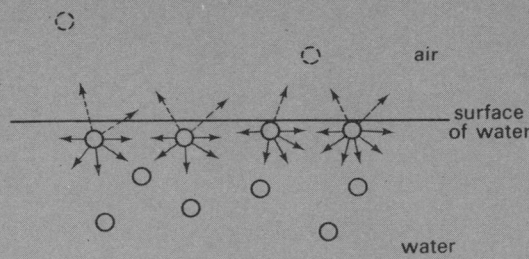

Fig. 33 The net resultant force on each molecule will be a force directed into the liquid perpendicular to the surface

Figure 33 illustrates the interfacial forces that might exist at a surface between, for example, water and air. There will be a net resultant force acting on the water molecules at the surface, directed into the liquid. It is not possible to detect this force directly in a normal container of liquid, but it will also give rise to an excess pressure inside a drop or bubble, and this pressure can be measured. An apparatus designed for making such measurements is illustrated in Fig. 34.

Because there is a force acting inwards on the surface of a drop, any increase in the size of the drop will require work to be done against this force. The force is proportional to the number of molecules on the surface and hence to the area of the surface. The more fundamental property of the surface is its potential energy, known in this case as *surface energy*. In any system, potential energy always tends to a minimum value, and because surface energy is related to surface area, a liquid/gas or liquid/solid interface of high surface energy will tend to cause the liquid to have as little area as possible. Thus water in air will form spherical drops because the surface energy between water and air is high. The surface energy

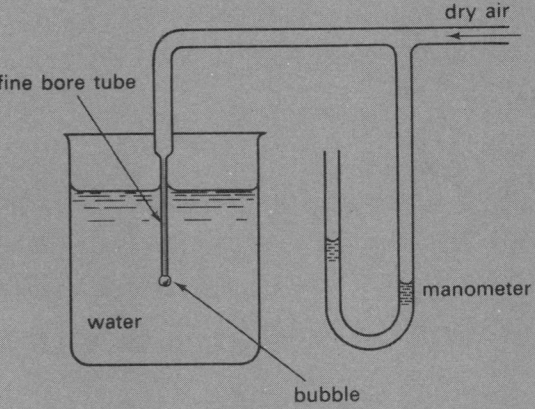

Fig. 34 Apparatus for measuring the excess pressure inside an air bubble

of critical importance in relation to the viscosity of the lubricant, if control is to be effective. Accuracy of alignment is essential since any eccentricity between the seal parts may set up a pumping action with the reverse effect to that required! Labyrinth seals are widely used in some power stations for grease-lubricated bearings on coal conveying plant. These can be made from a plastic material which has greater flexibility in service, high corrosion resistance, and low manufacturing and assembly costs.

The annular groove seal consists of a series of circumferential grooves located in the bore of the bearing housing through which the shaft passes. It prevents the leakage of lubricant owing to capillary action. The grooves interrupt the capillary flow by providing an increase in the area of the oil/air interface (see text B, page 28). When used as a grease seal the grooves are grease-filled, forming a totally enclosed bearing.

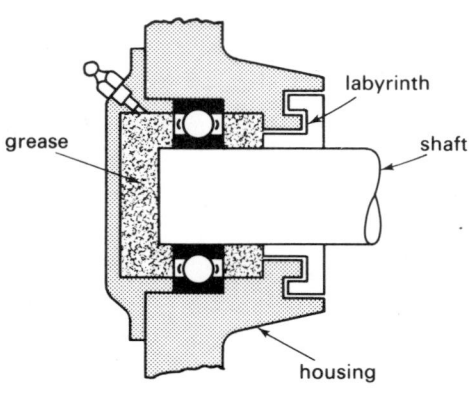
(a) labyrinth seal

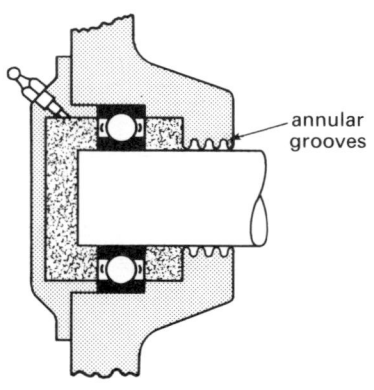
(b) annular groove seal

Fig. 35 Non-contact seals

---

between water and glass is low and water will therefore spread over the glass surface and 'wet' it. Lubricants, fluxes and adhesives must all have low surface energies when in contact with solid surfaces in order to 'wet' them. Thus a fibrous material will retain a lubricant in its pores and form a seal by virtue of the large surface area of the fibres. The liquid would run out and form drops if its surface energy were high.

In order to obtain an expression for the surface energy of a spherical drop let us assume that the resultant surface force per unit area is P. This must be equal to the difference in pressure between the inside and outside of the drop required to balance the surface force; a difference referred to as *excess pressure*. Suppose further that the radius of the drop is R and that it is increased by a small amount δR (see Fig. 36). The work done on each unit area of the surface is $P \delta R$, and since the total surface of the drop is $4\pi R^2$, the total work done is $4\pi P R^2 \delta R$. In the course of this process the increase in area of the surface (neglecting the term in δR^2) is:

$$4\pi (R + \delta R)^2 - 4\pi R^2 = 8\pi R \cdot \delta R \quad \ldots \quad (6)$$

Thus to create $8\pi R \, \delta R$ of surface, $4\pi P \cdot R^2 \delta R$ of energy is required. In other words the energy required to generate each unit area of surface is:

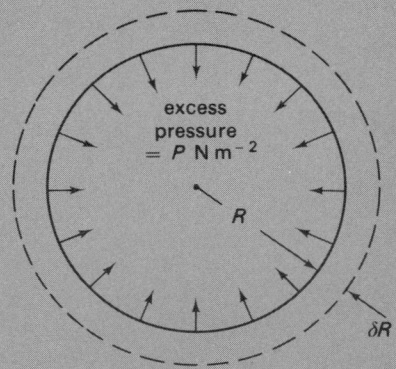

Fig. 36

$$\frac{4\pi P R^2 \delta R}{8\pi R \, \delta R} \quad \text{or} \quad \frac{PR}{2}.$$

Thus surface energy $\quad T = \dfrac{PR}{2} \quad \ldots \ldots \ldots \ldots (7)$

Note that for a bubble rather than a drop there are two surfaces and this equation must be applied separately to each surface to determine the total surface energy.

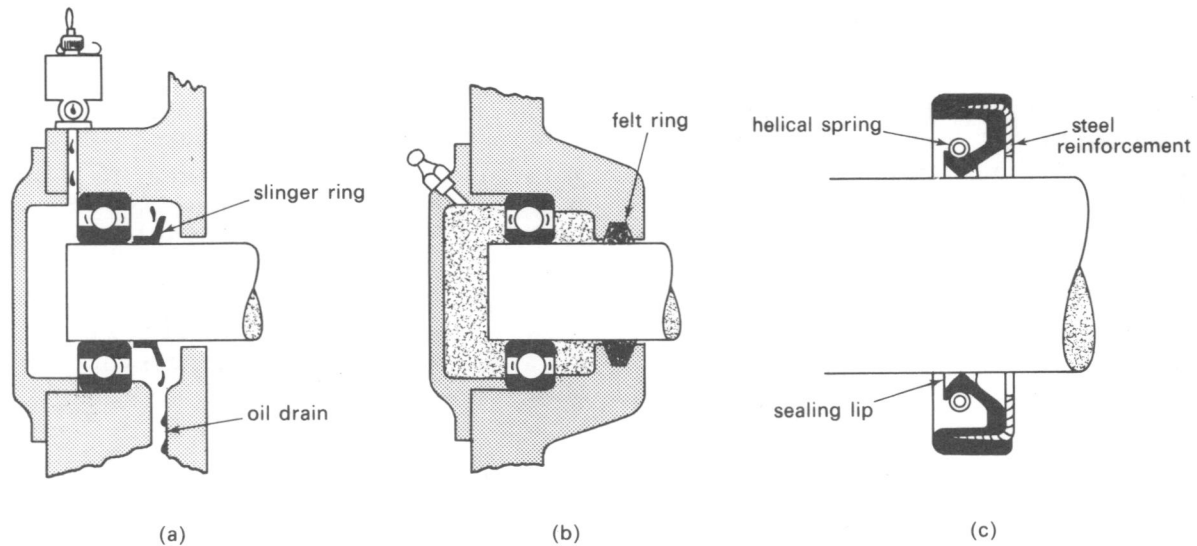

Fig. 37 (a) slinger, (b) felt and (c) lip seals

The spinner or slinger ring is fitted on oil-lubricated bearings and is essentially a flange or ridged collar secured to the shaft near the bearing enclosure. As the slinger rotates, oil is flung off the flange and gets no further along the shaft.

The simplest form of contact bearing is the *felt seal*. It comprises in most cases, a jointless ring of woven felt located in a groove in the bearing cover, and compressed against the spindle or shaft (see Fig. 37b). It is cheap, easy to install, and is effective under moderate conditions.

Lip seals make contact over a much narrower surface (see Fig. 37c), so that wear is negligible. The seals are manufactured as complete units consisting of a ring moulded from a suitable flexible material (e.g. synthetic rubber) and stiffened if necessary with a metal insert. The seal is held in contact with the shaft by a helical spring. This type of seal has great

Capillary action

Liquid will creep through a narrow space between two solid surfaces because there is a *small* area of liquid/air interface of high surface energy.

Consider what happens in the case of a fine-bore tube held vertically in a bowl of liquid (Fig. 38). An air/glass interface has a higher surface energy than a glass/water interface and the water tends to spread along the glass surface, giving rise to the meniscus commonly found when water is kept in a glass container. The pressure inside the liquid surface will be less than the pressure outside the surface, usually atmospheric, by approximately $2T/R$ where R is the radius of the curved liquid surface. (This follows from equation 7.) The liquid will move up the tube in order to attain a balance of pressure at X. To prevent movement of liquid in this way either the gap must be large enough to give sufficient high-energy surface area to prevent appreciable capillary movement, or so small that the resistance to flow of the liquid effectively prevents capillary creep. Thus a labyrinth seal must have a sufficiently wide channel to prevent significant capillary movement whilst a simple lip seal must grip sufficiently tightly to leave no gap.

The surface energy, which is the energy required to extend a liquid surface by unit area, is the same whatever the shape of the surface. Thus the surface energy does not have to be measured at the point of application but can be measured in any suitably devised piece of apparatus. The wire-frame method is commonly used in elementary work; it measures the force which results from the existence of surface potential energy. It depends simply on measuring, by means of a balance, the force necessary to remove a wire frame from a liquid, both with and without a liquid film within it (Fig. 40).

If a liquid film is present, then additional work must be done in lifting the frame from the liquid. If the frame is x metres wide, and is lifted y metres, then

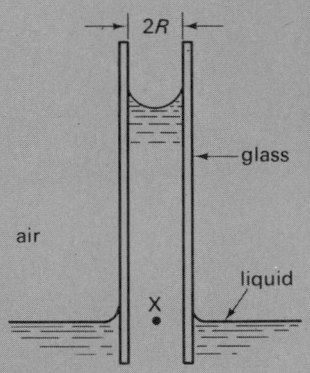

Fig. 38

flexibility both axially and radially, and is frequently termed an automatic seal; any pressure increase in the housing enhances lip contact.

Mechanical seals are basically different from other types in that the contact surfaces are at right angles to the axis of rotation, and in their simplest form consist of a collar attached to the shaft pressing against the bearing housing or cover, in much the same manner as a thrust bearing. In practice, however, the contact faces are usually separate rings of a different material from the shaft or housing (carbon is commonly used for one of the rings) and are thus easily replaced when worn. To cater for axial movement and shaft misalignment the rotating sealing ring may be held in contact with the mating surface by a compression spring as shown in Fig. 39, and this also automatically takes up any wear at the contact face. Mechanical seals, because of their more complex nature, are comparatively expensive and are used where exceptional conditions warrant the higher costs. They are often used as gland seals to prevent the leakage of gas or liquid from a pressure casing into an adjacent bearing. Frictional losses are low but it is important

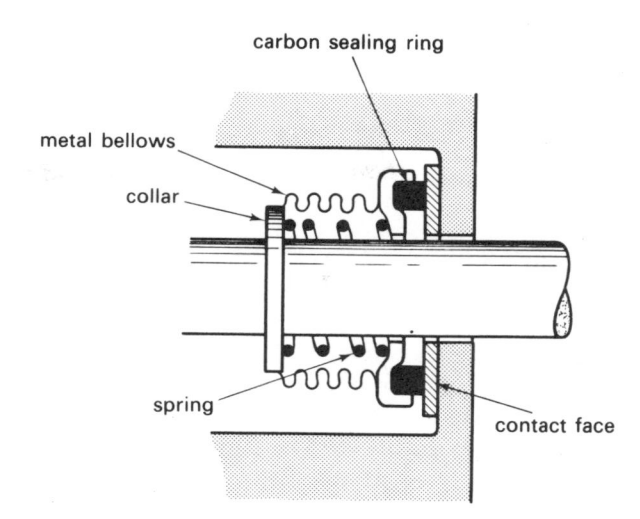

Fig. 39 Mechanical seal

that the contacting surfaces are smooth.

Since they are frictionless, non-contact seals are used in high speed machines. They do not wear and consequently require no adjustment or replacement. If they are to work successfully, felt and lip seals require a fine surface finish on the shaft on which the seal runs. Sometimes combinations of the above types of seal are incorporated in bearing enclosures, for example, annular grooves with a felt ring.

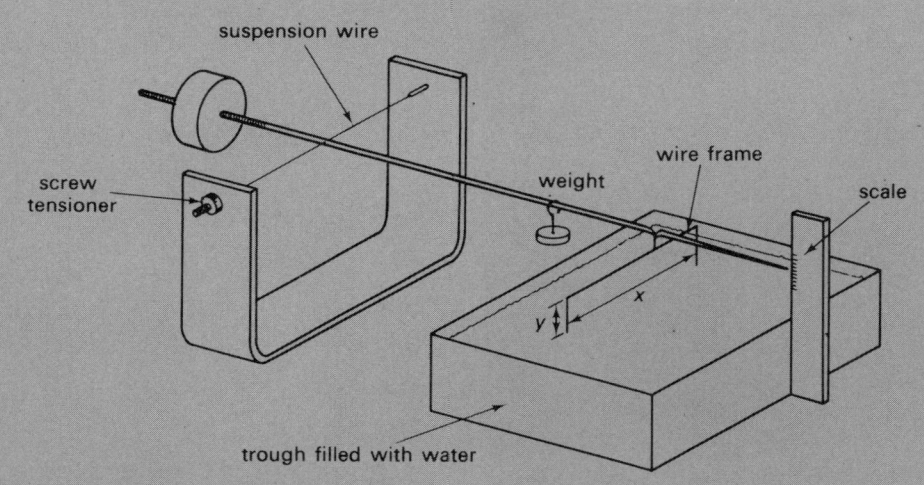

Fig. 40 Searle's apparatus for measuring surface tension

the extra work necessary is $2xyT$ (where T is the surface energy, and the coefficient 2 comes from the fact there are two surfaces). This work is done by the balance in exerting a force which lifts the frame by y metres. The force is therefore $2Tx$. T is therefore not only the value of the surface energy, but is also the force exerted per metre of surface. The force has a direction which lies along the surface, as in the experiment, and is known as the *surface tension*. It may seem strange that a force which originates as a net force directed into the liquid should give rise to a surface tension along the surface and perpendicular to the molecular resultant force. Surface tension is an expression of the force required to prevent the reduction of the area and a consequent reduction in surface energy.

Surface energy values range typically up to 3×10^{-1} joules metre^{-2}. The surface energy of an air/water surface is 7×10^{-2} J m^{-2}; thus the surface tension in a water film is 7×10^{-2} N m^{-1}; adding a detergent to the water may reduce this by up to 60 per cent.

4 The nature of surfaces

Recently, industry has paid increasing attention to the roughness of bearing surfaces and its effect on the life and efficiency of machines and mechanisms. It is the machining processes during manufacture which give bearing surfaces their texture; it will, therefore, be worth discussing surface finish and how its qualities are measured and specified.

No solid surface is ever perfectly flat: if highly magnified any surface would be seen to be made up of peaks and valleys, the height and spacing of which depends upon the method used to machine the surface (see Fig. 41). A finely ground surface will, obviously, be smoother in texture than a coarsely milled one. Methods of measuring the degree of smoothness of a surface will now be discussed.

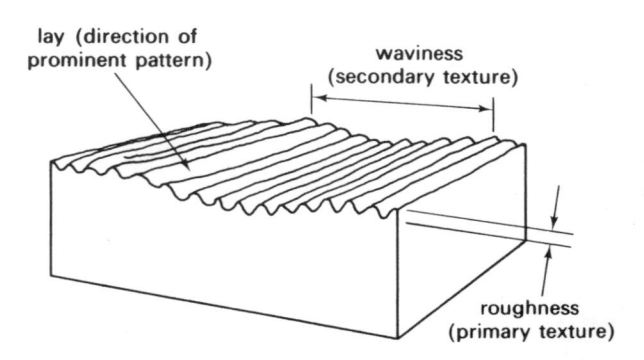

Fig. 41 Characteristics of a machined surface

Centre-line average (CLA)

The texture of a machined surface is shown, highly exaggerated, in Fig. 41. The irregularities of the surface are divided into two grades and are defined in BS.1134: 1961 as *roughness* (or primary texture) and *waviness* (or secondary texture). Roughness is a result of the action of the cutting tool during machining. Waviness, on the other hand, is dimensionally larger than roughness and is caused by the combined effects of machine or work deflection and vibration. The direction of the texture pattern is termed the *lay*. Because of the three-dimensional character of the surface irregularities, a comprehensive measurement is virtually impossible. In practice, measurements are usually made of the profiles of plane sections taken at right angles to the lay.

The degree of finish of a surface is given as a centre-line average number, a dimension which is now internationally accepted. It is in fact a measure of an average height, and depends on the graphical representation of the surface.

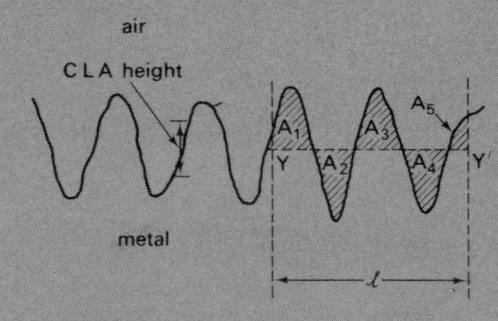

Fig. 42 Determination of CLA (centre-line average) from a trace

To determine the CLA value from a trace of the surface a line YY' must be drawn through the sampling length ℓ such that the total area above YY' is equal to the total area below YY' (refer to Fig. 42). In other words $A_1 + A_3 + A_5 = A_2 + A_4$. The CLA height of the trace is given by

$$\text{CLA}_{\text{trace}} = \frac{A_1 + A_2 + A_3 + A_4 + A_5}{\ell}.$$

The trace is a graphical representation of the surface

Figure 43 shows part of a stylus-type instrument used for producing a graph of the surface being checked. The sharply pointed diamond stylus is drawn lightly across the surface and travels up and down the surface irregularities relative to the skid, which acts as a datum. The stylus movement is converted into a varying electric current which is amplified and fed into a recorder, producing a trace similar to that shown in Fig. 42. For measuring highly polished surfaces where the irregularities are so small that the probe cannot discriminate between them, a microscope incorporating an optical device called an interferometer can be used to give an indication of the degree of surface roughness.

The cheapest and easiest method of measuring finish is simply to look at the surface and 'feel' it by scratching a finger nail across it. The look and feel of the surface can be compared with standard surface-finish references. These standards are produced for most methods of metal cutting; this is essential, as different cutting processes give different finishes. The standards are formed from stainless steel and are graded according to the centre-line average (see lower text, below and opposite), and the particular method of cutting. It may seem strange that this procedure, which depends entirely on subjective judgement, is considered to be adequate for general purposes.

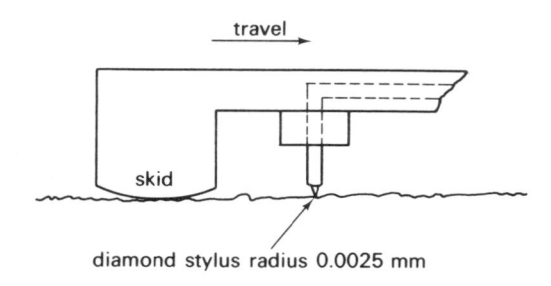

Fig. 43

with a very high vertical magnification, and to determine the CLA of the *surface* this must be taken into account.

$$\therefore CLA_{surface} = \frac{A_1 + A_2 + \ldots + A_5}{\ell} \times \frac{10^6}{\text{vertical magnification}}.$$

If the dimensions of the graph are expressed in metres, the CLA value calculated from the above equation is given in microns (1 micron = 10^{-6} m). A few typical values for surface roughness are given in Table 4.

TABLE 4

CLA values for different machining processes

Machining process	CLA m × 10^{-6}
Turned	1.5 – 10
Face milled	1.0 – 8
Coarse ground	0.3 – 0.5
Polished	0.05 – 0.2

Hardness

It is important, when considering the characteristics of a surface, to take into account its hardness, especially as the wear of metals in sliding contact may depend upon the relative hardnesses of the surfaces (see pages 11 and 12). Hardness can be defined as the resistance of a material to penetration of its surface. The surface hardness of certain substances, for example sintered materials, cannot be modified to any great extent, whereas with many alloys, particularly ferrous alloys, the surface hardness can be varied by heat treatment to give a precise value.

With certain equipment parts, e.g. coal handling plant in a power station or coal mine, high hardness is the most important property required in the components, mainly to resist wear. In the manufacture of these parts a minimum surface hardness value may be specified by the customer and, before acceptance, this may require to be checked by a hardness test. A hardened steel ball or a diamond point is impressed on the prepared surface of the metal for a limited period

Rough or smooth?

From experience you may think that the frictional force between smooth surfaces in moving contact is lower than that for rough surfaces, but this is not necessarily true. We have seen that the force of friction is the force needed to shear junctions that form over the real or actual area of contact between two surfaces. If clean surfaces of mica are carefully placed together adhesion occurs and the coefficient of friction is found to be infinite. This is because surfaces of mica are molecularly smooth over relatively large areas. It has also been found that some fairly rough surfaces have shown very low coefficients of friction (diamond is an example). This would indicate that a smooth surface does not necessarily have a low coefficient of friction, and conversely, that frictional forces between rough surfaces are not necessarily high.

In general, however, the smoother the bearing surface, the lower the ensuing wear. In engineering applications, rough surfaces (say, from machining) wear rapidly at first because the peaks are so large that they penetrate the normal lubricating film, and metal-to-metal contact occurs with corresponding wear. Deformation of the surfaces can also occur in that the 'peaks' are pushed into the 'valleys'. When the peaks are worn away, the rate of wear decreases considerably, mainly because the normal lubricant film keeps the surfaces separate. This initial wear process is commonly known as 'running-in', and while metal-to-metal contact is occurring, care should be taken to see that the bearing surfaces do not reach such a high relative velocity that excessive temperatures are reached. Any bearings manufactured with rough surfaces, may, after only a short time, become too 'sloppy' for use. Surfaces are therefore usually machined as smoothly and accurately as possible. (Can you think of examples where accurate, smooth bearings are vital?)

under a given load. Depending on the 'hardness' or 'softness' of the surface the indentation will vary in area; the area of the impression is measured and on the basis of load sustained per unit area, a hardness value can be calculated. Two hardness tests are in common use, the Brinell and the Vickers Diamond Pyramid, the hardness value being stated numerically in both systems. During manufacture, samples are usually provided for laboratory tests as part of the quality control procedure. In the power station a heavy component may have to be hardness-tested *in situ* and this can be conveniently done by means of the Shore scleroscope. The scleroscope is an instrument which measures the hardness related to the elastic properties of the material. A diamond-tipped weight is dropped through a glass tube from a height of 25 cm on to the surface to be tested. The rebound of the weight after striking the surface is measured and recorded; the harder the material, the higher the rebound. Several readings are taken, the average indicating the hardness of the work piece (the surface finish of the work affects the instrument readings to some extent).

At first sight you may think that with hard metals in contact, the deformation of the asperities will not be large, giving rise to a small area of real contact to be sheared, and hence a low coefficient of friction. But the shear strength for hard metals is higher than that for soft metals. This means that although the area of contact at the interface of soft metals, (or a soft metal in contact with a hard metal) is large, the shear strength is low (i.e. a low force is needed to shear the junctions). Consequently the coefficients of friction for soft and hard metals are not very different.

5 The selection of lubricants

A good maintenance engineer, in spite of being cost-conscious at all times, will not attempt to cut costs on the quality of the lubricant purchased for a particular item of plant. He will try to obtain the best possible lubricant, knowing that any extra cost incurred will be recovered many times over in the extended periods between servicing the plant, and in the resulting reductions in labour and material costs. The *outage* time, or the time a certain section of the plant is out of service, will be reduced, with consequent improved availability of the plant (i.e. the plant is available for operation for longer periods). The right lubricant, in the right place at the right time, is vital if a power station or any other modern industrial plant is to achieve a high operational efficiency.

It is the responsibility of the engineer to keep the machines in good order and to see that they are correctly lubricated. He will generally accept the recommendations of the supplier of the equipment as to the correct lubricant to use. This may sound obvious, but think of the complications in a power station, where many types of machines and mechanisms (electric motors, pumps, compressors, fans, control systems, turbines, valves, dampers, conveyors) each have their own specified lubricant. It may mean that as many as sixty different grades of lubricant will be stored at the power station ready to meet the requirements of maintenance schedules. To help prevent incorrect lubrication of the different items of plant, a code has been developed using differently shaped and coloured symbols to represent the type of lubricant and the frequency of application, e.g. a red triangle means a light lubricating oil, weekly. The symbol is simply painted on the machine near the point of application of the lubricant. You get an idea of the quantities of oil used in a power station when you consider that a 500 MW turbine generator requires about 4 000 gallons of high grade lubricating oil every minute, or 5 760 000 gallons per day! The choice of lubricant depends mainly on the way

Viscosity

An important property of a fluid lubricant is its viscosity, which in simplest terms is the resistance of a fluid to flow. We saw in Chapter Two that a lubricant is a material which is continually sheared when interposed between two surfaces in moving contact. By virtue of its viscosity a fluid offers a certain resistance to the force tending to shear it, and it is this resistance which is the force of friction between the lubricated surfaces. It is an internal friction effect of the fluid.

One of the first scientists to study the problems of fluid motion was Sir Isaac Newton, who observed that an unconfined liquid or gas would flow when a force was applied to it. He found that there was a resistance to flow and ascribed this to 'a lack of slipperiness of the parts of the liquid'.

We will first of all see what happens in a simple case of fluid shear. Let us suppose that a flat plate moving with a velocity U is separated from a stationary surface by a film of fluid lubricant of thickness d (see Fig. 44). The stationary surface is flat and parallel to the lower surface of the plate and we must assume that the plate is so light that the fluid is under negligible

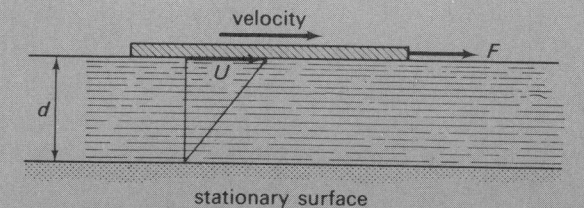

Fig. 44 Shearing a viscous oil film

the lubricant behaves in the particular bearing considered. You do not lubricate your watch with car steering-joint grease (let's hope!), because steering-joint grease is too thick, or heavy, and would just clog up the works.

Viscosity

From our understanding of the chemistry of oils we know that all mineral oils (which constitute the largest group of lubricants) are of the same hydrocarbon group. They vary principally in a property known as viscosity, which is a measure of how 'thick' or 'thin' an oil is. Viscosity is the resistance of a fluid to flow. Fluids such as honey and treacle have a high resistance to flow: they would require, for instance, a large force to pump them through a narrow pipe, and are said to have a high viscosity. Water and petrol, on the other hand, have low viscosities and flow easily. It is stated below that viscosity is the resistance of a fluid to shear, and as this shearing takes place inside the fluid and not at the boundary between a fluid and a solid surface, shearing of a fluid can be considered as an internal frictional effect (fluid sliding on fluid): viscosity is the measure of this internal friction.

The viscosity of a liquid can be considered comparable to the mechanical shear strength of a solid material. It follows that viscosity is that property of a fluid which resists mechanical force; this in turn determines the frictional resistance of a lubricant, the heat generated in the lubricant film, and the load that the bearing will carry.

With nearly all liquids an increase in temperature results in a reduction of the cohesive forces between the molecules of the liquid. The force needed to shear the fluid internally is lowered, and hence the viscosity decreases. It is important, therefore, when specifying an oil for a particular bearing that its viscosity is such that it is a satisfactory lubricant at the normal working temperature of the bearing. A change of only a few degrees in temperature can mean a significant change in viscosity. (For example, the viscosity of water at 50°C is about half its viscosity at 20°C.) Figure 45 shows a typical viscosity-temperature curve for a light lubricating oil over a temperature range.

pressure. The fluid can be thought of as being made up of a number of thin parallel layers. The layer nearest the moving surface adheres to it by forces of molecular cohesion, and therefore travels with a velocity U. Similarly the fluid layer in contact with the stationary plate has zero velocity. Successive layers between the surfaces will slide over each other under the action of the force F applied to the moving plate, and their velocities will vary as shown in the figure. (Try shearing a pack of cards in the same manner.)

The variation of velocity with the distance from the stationary surface is a straight line. The force required to slide the layers over each other is a shearing force: it is the force needed to shear the forces of attraction between fluid molecules. The applied force F, needed to maintain the plate at a uniform velocity U, is a measure of how viscous the fluid is, i.e. how easily it is sheared: if the force F is large the viscosity or internal friction of the fluid is high. Each layer of fluid will be subjected to a shearing stress τ, given by

$$\tau = \frac{\text{force}}{\text{area}}$$
$$= \frac{F}{A}$$

where A is the area of the lower surface of the plate.

Experimental results show that the shear stress τ is proportional to the *velocity gradient* or the variation of velocity with distance from the stationary surface. Hence

$$\tau \propto \frac{U}{d}$$

$$\therefore \quad \frac{F}{A} \propto \frac{U}{d}$$

or

$$\frac{F}{A} = \eta \frac{U}{d}.$$

The constant of proportionality, given the symbol η, is called the absolute (or dynamic) viscosity of the fluid. For practical purposes nearly all fluid lubricants conform to the above equation. It follows that:

$$\tau = \eta \frac{U}{d} \quad \ldots \ldots \ldots \ldots (8)$$

The units of dynamic viscosity are those of stress × (velocity gradient)$^{-1}$.

$$\eta = \frac{\text{force}}{\text{area}} \times \frac{\text{distance}}{\text{velocity}}$$
$$= \frac{F}{A} \frac{d}{U}$$

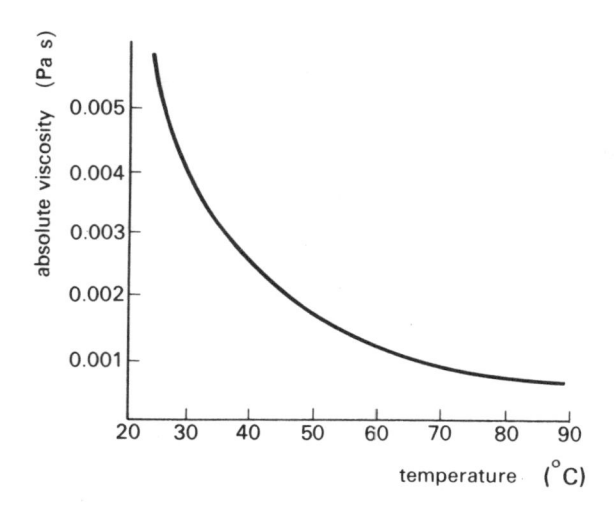

Fig. 45 Variation of oil viscosity with temperature

The temperature variation of viscosity means, of course, that an oil which is satisfactory at low temperatures may become too thin to provide adequate lubrication when, say, an engine has warmed up. An oil which has a small variation of viscosity with temperature is required in such cases and 'multi-grade' oils have been introduced to overcome this difficulty, especially for car engines. They are hydrocarbon oils with special additives which provide satisfactory lubrication in summer weather as well as during the winter.

The units of viscosity

Viscosity is defined as shear stress divided by velocity gradient and has the units $N\,m^{-2}\,s$, also known as Pascal seconds, or Pa s (see text B below). Sometimes it is more useful to know the ratio of the absolute viscosity of a fluid to its density. This ratio is termed the kinematic viscosity and has the dimensions $m^2\,s^{-1}$. More often than not, figures for the viscosities of oils are used for comparison purposes and it is not necessary to know the viscosities in Pa s. It is easier to conduct a test with units of $\dfrac{N}{m^2}\,\dfrac{m}{m\,s^{-1}}$ or $N\,s\,m^{-2}$.

Dynamic viscosity is usually expressed in pascal seconds or Pa s where a pascal has the dimensions $N\,m^{-2}$.

In some cases in the solution of problems in the study of fluid flow the ratio of absolute viscosity to density is required. This ratio is termed kinematic viscosity, and is given the symbol ν.

$$\nu = \frac{\eta}{\rho}$$

with units $\dfrac{N\,s}{m^2}\,\dfrac{m^3}{kg}$ or, rearranged, $N\,\dfrac{s\,m^3}{m^2\,kg}$.

Now the units of force (N) are those of mass × acceleration or $kg\,m\,s^{-2}$. Hence units of

$$\nu = \frac{kg\,m}{s^2}\,\frac{s\,m^3}{m^2\,kg} = m^2\,s^{-1}$$

hence kinematic viscosity has units of area × (time)$^{-1}$.

The concentric oil bearing

The first time this theory of viscosity was put to practical use was in 1883 when the Russian Petroff investigated 'Friction in machines and the effect of the lubricant'. He produced an equation which will accurately predict the bearing friction due to viscous shear in the oil film of a plain bearing where the shaft and housing are concentric. Concentricity is established when the speed of rotation is high and the applied load is light, and many bearings do in fact operate under these conditions. Normally the clearance is small compared to the shaft radius and the bearing surfaces may be considered as if they were parallel and equal in area. We must assume that there is negligible end-leakage of oil from the bearing.

using, for example, the efflux-type apparatus such as the Redwood viscometer. The result of the test is expressed as 'seconds Redwood': this is the time taken for the oil under test to pass through an accurately dimensioned orifice. The longer the time, the more viscous the oil. Turbine oils in common use range in viscosity from 135—465 seconds Redwood 1, which in terms of kinematic viscosity is a range of from 5×10^{-5} m² s⁻¹ to 20×10^{-5} m² s⁻¹.

You may be aware of another grading of viscosity, that of a number sometimes followed by the letter W. This is a classification used for engine and transmission oils, which was introduced by the American Society of Automotive Engineers. The higher the number, the greater the viscosity. A letter W after the number expresses a winter grade oil. The range for engine crank case oils is 5W, 10W, 20W, 20, 30, 40 and 50.

The measurement of viscosity (Viscometry)

Because viscosity is the most important physical property of a lubricating oil a great deal of attention is devoted to its accurate determination. Viscosity-measuring devices are known as viscometers, and a number of different types are in use. The most common are the capillary, efflux, rotational, and falling sphere viscometers. In this book we will discuss only the first two types.

Capillary viscometers

The capillary viscometer accurately measures the kinematic viscosity of oils. The oil under test is allowed to flow through a capillary tube of known dimensions. As viscosity varies with temperature a thermostatically controlled water bath surrounds the viscometer tube. The time of flow through the capillary is related to the kinematic viscosity of the oil. To convert to absolute viscosity, an accurate density-temperature relationship for the oil must be known.

Efflux viscometers

The above process is laborious and the

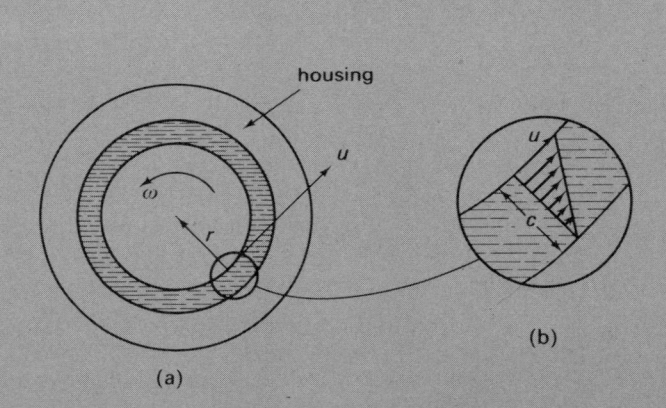

Fig. 46 The concentric oil bearing

Figure 46 shows a fully lubricated journal bearing of length ℓ with a concentric shaft of radius r, rotating with an angular velocity ω. The clearance or thickness of the oil film is c and the velocity of the surface of the shaft is u. The velocity gradient across the oil film is constant and equals $\frac{u}{c}$ (see Fig. 46b).

From equation 8

shear stress = viscosity × velocity gradient

i.e. $\tau = \eta \frac{u}{c}$

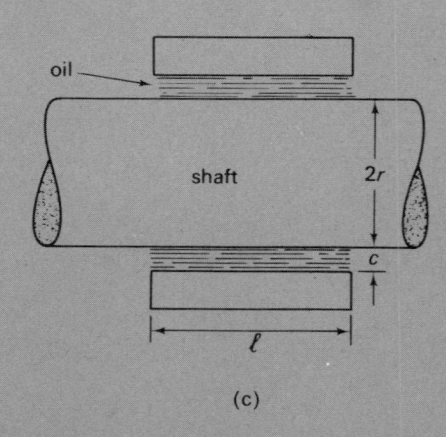

where τ, the shear stress in the oil film, is given by

$$\tau = \frac{\text{shear force}}{\text{area}} = \frac{F}{2\pi r \ell}$$

hence $\frac{F}{2\pi r \ell} = \eta \frac{u}{c}$

and $F = \frac{2\pi r \ell \eta u}{c}$

Now the velocity $u = r\omega$

apparatus is relatively fragile. In industry the Redwood viscometer is used to provide a rapid check on the viscosities of oils at specified temperatures. It simply measures the time taken for oil to flow through an orifice in the bottom of a vessel. The No. 1 instrument, used for measuring low and moderate viscosities, is shown in Fig. 47. The oil cup has a capacity of 50 cm³ and discharges through an accurately dimensioned agate jet. The time taken for the oil to flow through the jet is expressed as 'seconds Redwood', and this gives a measurement of the viscosity of the oil. There is no simple relationship between the flow time and viscosity, and conversion tables must be consulted if the kinematic viscosity is required. Here again, the absolute viscosity can be deduced if the density is known. Redwood viscometers are cheap, robust, easy to use, and provide for comparisons of oil viscosities at specified temperatures. Other efflux instruments, the Saybolt and Engler viscometers, are in use in the United States and Europe respectively.

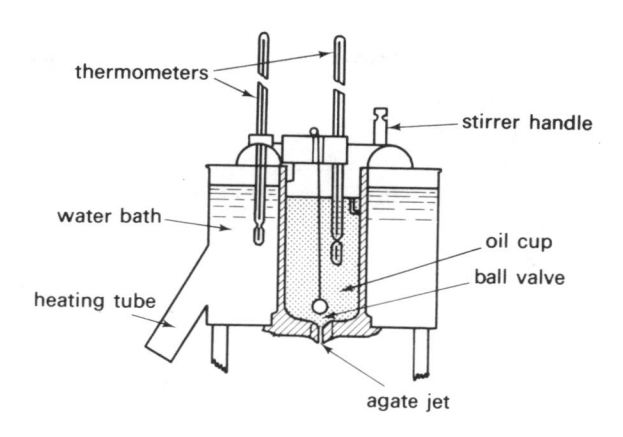

Fig. 47 Detail of Redwood 1 viscosity instrument

hence
$$F = \frac{2\pi r^2 \ell \eta \omega}{c}$$

and is the magnitude of the friction force in the oil film.

This viscous force acts at a distance r from the centre of rotation, hence the resisting torque T is given by

$$T = \text{force} \times \text{radius}$$

$$= \frac{2\pi r^3 \ell \eta \omega}{c} \quad \text{(Petroff's equation)}$$

and the power P, expended in shearing the oil film (which is a measure of the energy loss in the bearing) is given by

$$P = T \cdot \omega$$

$$P = \frac{2\pi r^3 \ell \omega^2 \eta}{c}.$$

As the load on the bearing increases, the shaft becomes displaced in the bearing and the friction force slowly increases. Under moderate operating conditions this increase is small and the Petroff equation will produce a sufficiently accurate estimate of the frictional torque. (Could you devise an apparatus for measuring the absolute coefficients of viscosity, η, of different oils using the above theory?)

Viscosity Index

Various methods have been used to express the viscosity-temperature relationship simply, but the method most commonly used is the Viscosity Index, VI. It is an empirical number which indicates the degree of variation in viscosity with change in temperature and is important where the oil is to be used in situations where the operating temperature changes considerably. A high VI indicates little change in viscosity with temperature. The numerical value of the VI is established by comparing the variation in viscosity of the sample lubricant with that of two specific reference oils, one oil exhibiting a large change in viscosity over a specified temperature range, and the other showing a small variation. The former is given a VI of 0 and the latter a VI of 100. Each oil tested will normally have a VI occurring between these two extremes. The characteristic curve for a high index oil will be shallow, whereas the curve for a low index oil would be correspondingly steep. The temperatures normally used in the determination are 100°F and 210°F, and very often for commercial

Degradation

Lubricants deteriorate after a time owing to degradation, a breakdown of the hydrocarbon groups which is brought about by a number of factors, some of which are described below:

Oxidation Oxygen in the air reacts with some of the chemical compounds of which oils are composed, in a process known as oxidation. The rate of oxidation is increased at high temperatures and by the presence of moisture, dirt and metallic particles; in addition the cyclic variation of pressure and temperature inherent in recirculatory lubricating systems promotes this form of degradation. The products resulting from oxidation may be compounds that are either soluble or insoluble in the oil. Despite the oil purifiers normally found in recirculatory systems, the insoluble compounds settle out and may accumulate in quantities sufficient to interfere with the oil supply to the bearings. Soluble oxidation products are often slightly acidic and may cause corrosion of metal surfaces. Eventually, prolonged oxidative degradation causes the original physical and chemical properties of the oil to disappear completely. Oxidation can be reduced by proper bearing design in which aeration and temperatures are kept to a minimum and by the use of special additives in the oil which help to prevent oxidation.

Foaming The presence of air in a circulatory system is virtually unavoidable; the air enters through glands, joints and bearings, where slight negative pressures are set up in the vicinity of rotating parts. The air becomes closely mingled with the oil in the form of

and technical purposes the viscosity of an oil is quoted at these temperatures. (The Fahrenheit scale of temperature is still sometimes used by British Industry. The equivalent range in Celsius is from $38°C$ to $99°C$.) Referring to Fig. 48 the VI of the test oil has a viscosity index given by

$$VI = \frac{V_a - V_c}{V_a - V_b} \, 100.$$

Although the VI of a lubricant is governed by the type of hydrocarbon present, it can be modified by the addition of other chemical compounds to achieve special characteristics. In most circumstances it is advantageous to select a lubricant of high VI for the safety margin it affords in case of inadvertent temperature rise.

Hydrodynamic or fluid lubrication

In Chapter Two we saw that frictional wear is eliminated in a bearing in which the two surfaces are completely separated by a film of lubricant. The frictional force is the force required to shear the oil film, and is dependent on the viscosity of the lubricant.

At first sight it may seem impossible to prevent a heavy journal in a plain bearing from displacing all the oil under it, thereby causing metal-to-metal contact, even when the shaft is revolving.

In 1886 the Royal Society published the theory of hydrodynamic lubrication developed by Osborne Reynolds, who showed that pressures can be developed in the oil film which keep the surfaces apart as the journal rotates.

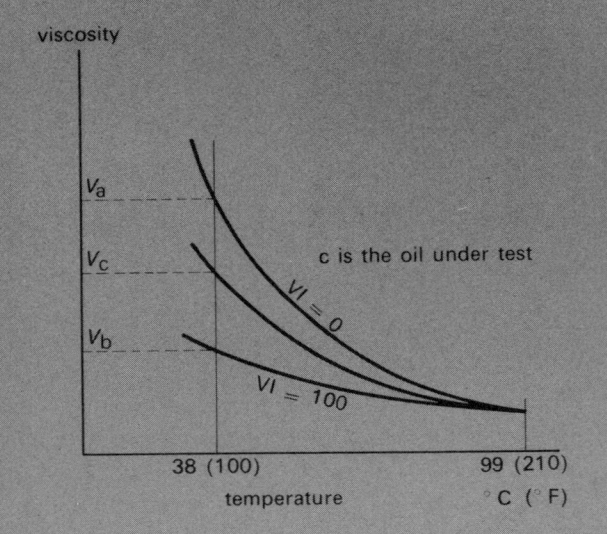

Fig. 48 Viscosity Index

tiny bubbles and is the cause of foaming, a condition which assists any oxidation by increasing the area of contact between air and the oil. The design of the oil system can do much to correct this fault by eliminating points of air entry, reducing excessive agitation in the oil returning to the sump, and by providing adequate air vents on reservoirs, casings, and drain tanks.

Heat In addition to the heat generated by internal friction in the oil film, the bearing may be heated by conduction from other parts of the machine, e.g. the steam casing on a turbine. The heat transferred in this way can be as much as twice the frictional heat and, if too intense, can reduce the viscosity of the oil film to a dangerously low value. The oil must therefore be capable of cooling, as well as lubricating the bearing. A turbine bearing may be designed to operate at temperatures approaching 70°C, which necessitates the use of high VI lubricant referred to on page 38, text B.

Water Contamination by water is common in circulating systems, the water originating from induced air and, in the case of the steam turbine, from steam leakages or faulty water-cooling systems. Combination labyrinth and slinger seals are fitted on the shaft near the bearing to combat contamination of the oil. If the oil is agitated it may emulsify but with oil in good condition the emulsion separates out and the water can eventually be removed by a centrifuge.

Rust, metallic particles, dirt, oxidation products and other impurities in combination form a sludge which resists oil circulation and are harmful if retained in the system.

We will now examine what actually happens to a journal in a lubricated bearing as it starts from rest. Figure 49a shows the journal at rest in the bearing; it has squeezed out the oil from directly beneath it. The clearance in the bearing has been exaggerated; in practice the clearance is usually about 0.001 times the shaft diameter. As the shaft begins to rotate, it tends to climb up the bearing surface due to the friction force between the surfaces (see Fig. 49b). A point is reached where the oil film drawn under the shaft causes the shaft to slip, and at a certain critical speed it takes up its equilibrium position as shown in Fig. 49c. The converging oil film causes an increase in the flow velocity as the oil passes between the surfaces and a hydrodynamic pressure is built up in this oil wedge. The pressure is great enough to keep the surfaces completely separated. Figure 49c also shows the pressure distribution in a journal bearing under hydrodynamic conditions. Each arrow is proportional in length to the pressure at that point.

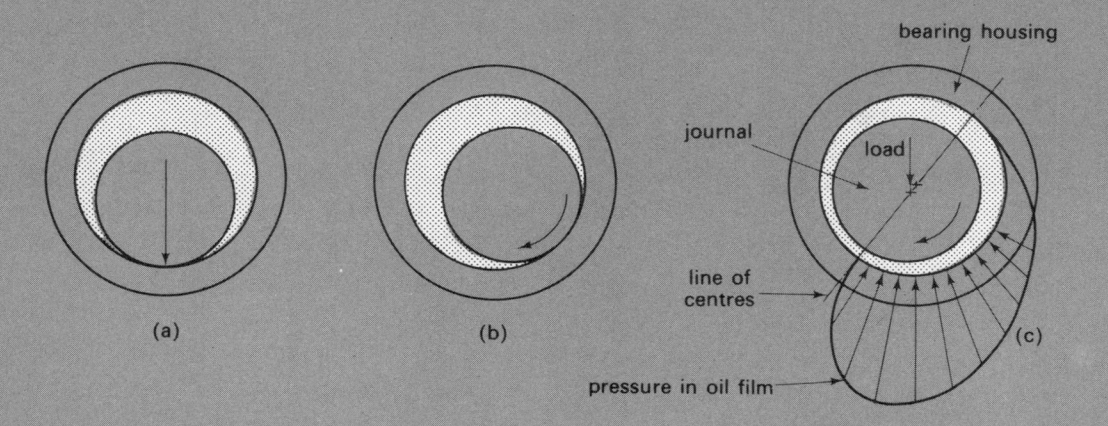

Fig. 49 Hydrodynamic lubrication

Consistency

The property of a grease that corresponds to the viscosity of an oil is its consistency. Consistency varies with temperature, but the variation has a far less noticeable effect than the thinning-out of an oil, and grease may be used over large temperature ranges.

Greases consist of mineral oils thickened with a metallic soap, and their consistency depends on the viscosity of the oil and the type of soap incorporated. Consistency is normally determined by measuring the depth to which a standardised weighted metal cone sinks into the grease over a specific period of time. A 'soft' grease will give a high number, a 'hard' grease a low one. As grease is sheared or 'worked' the consistency is reduced and the test is usually performed twice, for 'worked' and 'unworked' stages of the grease. If the difference in results is small the grease may be used for conditions that involve churning, although if the churning is too great, excessive temperatures will be generated and the grease will melt, and fail as a lubricant. All greases that are excessively sheared or churned during use will suffer a gradual diminution of consistency.

The choice of lubricants

The appropriate lubricant for a particular purpose must be chosen with care. Suitable lubricants should be chosen on the basis of the application itself. The ideal lubricant should first be determined and the bearing designed to accommodate it, rather than the other way round. What are the factors that influence the type of lubricant to be used? Why are some bearings lubricated with grease, some with oil, and others by solid lubricants such as graphite? The answer is not simple. The choice of lubricant is determined by

1. the operating speed
2. cost
3. operating temperature range
4. bearing housing limitations
5. ability to transfer heat
6. ability to protect the bearing (e.g. prevent corrosion)
7. methods of sealing
8. ease of application
9. degradation characteristics
10. the types of lubrication: hydrodynamic, boundary, extreme pressure, or hydrostatic.

Boundary lubrication

Fluid lubrication is not effective unless the lubricant film completely separates the surfaces. Under conditions of high loading or low speeds lubrication is by means of a *boundary* film, which may be only one or two molecular layers thick. The friction force depends on the chemical constituents of the lubricant and the nature of the surfaces, and is independent of viscosity. Some welded junctions form under boundary conditions, although there are far fewer than would exist with no lubrication, and frictional resistance is relatively low ($\mu = 0.02 - 0.1$). Boundary lubricants interact with the sliding surfaces to form very stable films. A good lubricant is molybdenum disulphide described in text B on page 23.

Extreme pressure lubrication

In certain gear trains, such as the hypoid type of motor car rear axle drive, high loads are combined with extremely severe sliding conditions. Also, the mating gears are usually both made of steel, and a lack of suitable lubricant would soon ruin the meshing surfaces. Special extreme pressure (EP) lubricants are available for use in such cases, as normal fluid or boundary lubricants would break down under the surface temperatures developed. EP additives are usually combinations of chemicals containing phosphorus, sulphur, or chlorine which become chemically active at high temperatures and combine with the surfaces to provide a film of low shear strength on the faces of the teeth.

Hydrostatic lubrication

In some bearings the boundary lubrication that takes place before a full fluid film is set up may be undesirable. Also, in parallel bearing surfaces such as slideways, no pressure can be generated in the oil film and lubrication will be largely of the boundary type. '*Hydrostatic*' bearings are designed to provide full fluid film lubrication in cases where boundary lubrication is undesirable, as in turbine alternator bearings and machine tool slideways. In the hydrostatic bearing, oil is fed under pressure to the bearing

It is difficult to make generalisations as to which lubricant and bearing to use for a particular job. It must be remembered that the choice of oil will chiefly be governed by its viscosity, and the choice of grease by its consistency, chemical stability and load-carrying capacity. Some plain mineral oils may have additives for special purposes, such as to prevent foaming or to keep particles of carbon in suspension in motor oils. Extreme pressure additives are used in gear oils to withstand the high loads on gear teeth surfaces.

The application of a grease depends on its chemical structure. Lime (or calcium) soap greases are relatively inexpensive, and are highly resistant to water. They serve widely for general purpose lubrication up to a top operating temperature of 50°C. Soda greases have a higher operating temperature (<120°C) but are soluble in water. Lithium greases are used extensively for general lubrication; they are rapidly supplanting both calcium and soda greases, as they combine the smooth texture of the former with the higher operating temperature of the latter. They operate up to 120°C and have satisfactory water-resistance.

Much research has been carried out during the last decade into the use of air and other gases as lubricants. The idea is not new: Hirn first mentioned air-lubricated bearings as long ago as 1854, but it is only in recent years that the idea has been applied to practical situations. Where are air bearings to be found? At present they are used mainly in space and aviation technology, and in high speed precision machines in the production engineering industry. (Precision grinding machines have to work accurately up to speeds of about 150 000 revolutions per minute.) When used for the bearings of high speed electric motors, the lubricating air can also be used to cool the rotor windings.

In cases where oil or grease would be unsuitable, such as at high operating temperatures or in processes where contamination cannot be risked (aerospace, medicine, and pharmacy) a 'dry' lubricant or bearing may be used. PTFE bearings are suitable for temperatures up to 200°C; nylon can be used at more moderate temperatures, but its coefficient of friction is greater than that for PTFE. Graphite and molybdenum disulphide give effective solid lubrication up to temperatures of 400°C. Carbon and graphite materials

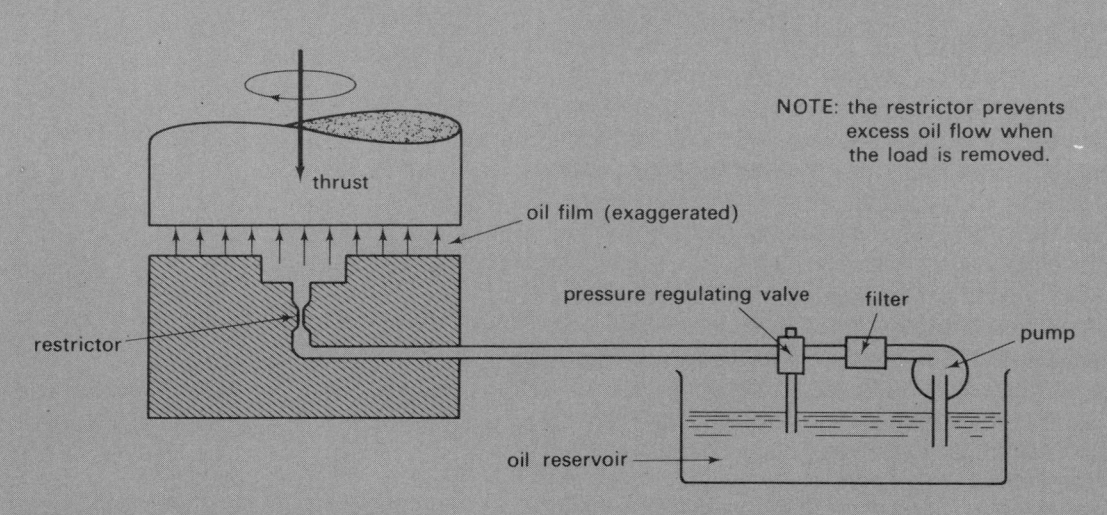

Fig. 50 Essential parts of a hydrostatic bearing

surfaces from an external source (Fig. 50). The pressure is enough to force the surfaces apart, and wear is virtually eliminated.

Hydrostatic bearings have the disadvantages, however, of a need for associated auxiliary equipment such as pumps, filters and pipe-work and considerable power must be expended in supplying oil under pressure to the bearing surfaces. Hydrodynamic bearings generate their own pressure and are cheaper and easier to install and operate.

are brittle, have low thermal conductivity and do not work well in very dry atmospheres or in vacua; they are usually incorporated in a metal matrix by impregnating a porous metal structure. But even these solid lubricants are not suitable to satisfy the repercussions of friction on the demands of modern engineering, which include high speeds, high temperatures and cryogenics. Investigations and experiments are being carried out at the moment in the whole field of solid lubricants, consisting of combinations of metals, carbons, polymers, oxides and ceramics, to answer these special requirements (especially the demands of the aerospace industry, e.g. for low friction in a vacuum). Because of their high cost, it is unlikely, however, that these exotic composites will ever be widely used in conventional engineering industries, but their development and adoption for special purposes is bound to increase during the next decade.

Gas lubrication

What properties commend the use of air as a lubricant? Gases are much less viscous than oils, so the frictional losses in an air bearing are reduced considerably, and there is consequently very little heating of the lubricant and little wear. Indeed, the high thermal resistance of air makes it ideal for operating at high temperatures, and because air does not suffer from the degradation effects of oils, it can be used in corrosive surroundings. And of course air is cheap! You may ask: why are air and gas bearings so uncommon? The answer is simple: air has a low load-carrying ability and gives no boundary lubrication at starting and stopping. It is useful as a lubricant, therefore, generally where loads are low and speeds are high, as in high-speed grinding machines. Solid lubricants such as molybdenum disulphide on the bearing surfaces can help reduce starting and stopping friction and wear; the alternative is to supply air under pressure to the bearing to keep the surfaces separated. Once rotating at a high enough speed the bearing can be self-acting (or 'aerodynamic', cf. hydrodynamic lubrication), drawing the air or gas directly from the surroundings. For high loads and low speeds compressed air must be constantly supplied to the bearing in order to keep the surfaces apart. Such bearings are termed 'aerostatic'. (In fact a hovercraft is simply a large aerostatic bearing.)

6 Friction devices

You may have the impression that friction causes a complete waste of energy, time and money, and that the millions of pounds spent on reducing the effects of friction in machinery would be saved if friction did not exist. But if this were the case much more would be spent on *providing* friction where it is needed, for friction is fundamentally necessary and many devices function solely because of the effects of friction. Buildings, for example, rely almost totally on friction to prevent them from collapsing (see *Structures*, Chapter Two). We have already mentioned briefly in Chapter One some of the places in a power station where friction is useful, such as in the pulverising fuel mills, where we are prepared to pay for the wear this useful friction causes, though if the wear can be kept to a minimum, so much the better. In this chapter we will discuss the operation of a few common friction devices.

There are plenty of familiar examples of the usefulness of friction; for instance, if it were not for the friction force acting at the four small areas of contact between the tyres of a car and the ground then the car would be totally uncontrollable — in fact it would not even be able to move! We depend on friction acting at these areas of contact under *all* road conditions. The tyres must be designed to prevent the build-up of water between tyres and ground in wet weather; this water would act as a lubricant and cause 'aquaplaning', i.e. the tyre moves on a cushion of water and all control is lost. To reduce the likelihood of aquaplaning a zig-zag pattern is incorporated in the surface of the tyre to displace rainwater to the rear of the tyre (at the rate of more than one gallon per second at sixty miles per hour for an average tyre). Tyres are a good example of designing for maximum friction over a range of surface-to-surface conditions.

The conveyor belt

Although conveyor belts are often used on a level, in power stations they are sometimes required to raise coal, and are therefore inclined at an angle to the horizontal. It is clear that a steep belt will be shorter than one at a low gradient, hence it will be cheaper and require less driving power. If, however, the belt is too steep the coal will continually roll or slide back down the belt. (Much depends of course on the relative position of the coal store and the boiler house.) What criteria govern the optimum angle of inclination of the belt, so that sliding or rolling just does not occur under all foreseeable conditions of coal condition, size and bulk density?

Consider the body resting without slipping on an inclined plane as shown in Fig. 51a. The friction force F resists the component $mg \sin \theta$ of the weight of the body along the plane, and the normal reaction R is equal to $mg \cos \theta$. θ is the angle of inclination of the plane. In text B, page 10, it was shown that there is a limit to the angle of friction, and that sliding will just occur when the **tangent of the angle of friction** ϕ is equal to the coefficient of friction μ between two

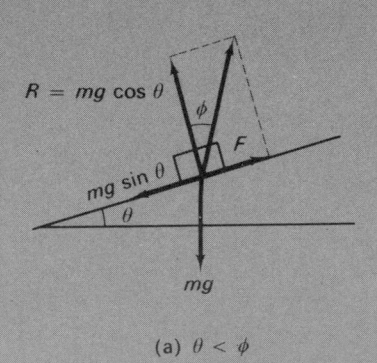

(a) $\theta < \phi$

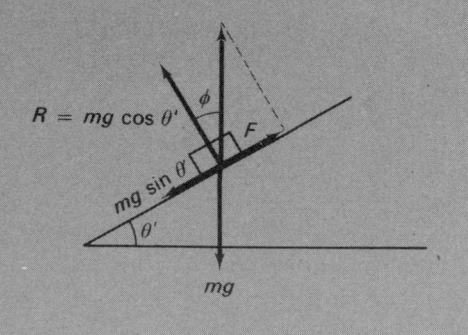

(b) $\theta' = \phi$

Fig. 51 Body on an inclined plane

Conveyor belts

On page 3 we mentioned the conveyor belt as a means of transporting material from the coal store to the bunkers. It consists essentially of an endless belt, one metre or more wide, and usually made of rubber reinforced with fabric or steel, stretched over two steel drums as shown in Fig. 52. The drive, often an electric motor and gearbox, is situated at the discharge end of the belt, and the free roller at the tail end (the feed end) is usually adjustable so that the tension in the belt can be altered.

To prevent the belt from sagging under its load, idlers are spaced at frequent intervals along the 'tight' side, or the loaded side. By arranging the idlers as shown in Fig. 54, and a flexible rubber belt, a trough can be formed which will allow more coal to be carried per width of belt and which will reduce the risk of coal spillage. There is no need for the empty 'slack' or return side of the belt to be troughed, and it therefore remains flat; the idlers supporting the return side are further apart.

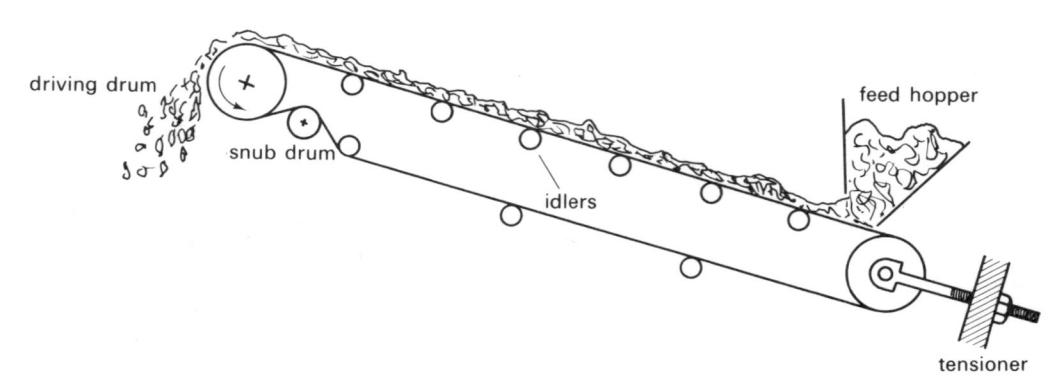

Fig. 52 A simple conveyor belt

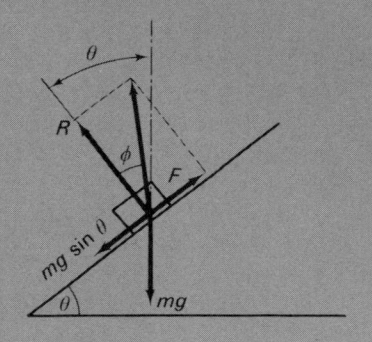

Fig. 53 $\theta > \phi$

surfaces. If the slope θ in Fig. 51a is increased, it will eventually reach a slope θ' such that $\theta' = \phi$ and the block will be just on the point of sliding (Fig. 51b). If the inclination is further increased (Fig. 53), the friction force F is insufficient to overcome the component $mg \sin \theta$ down the slope, and there will be a resultant force $mg \sin \theta - F$ tending to accelerate the body down the slope.

The body will just start to slide down the slope when the angle of inclination θ is equal to the angle of friction ϕ for the two surfaces in contact. The body will not slide down the slope if the angle of friction is less than the slope angle. With coal on rubber conveyor belts the maximum slope is limited to about 18°, whereas for wet ashes the slope may safely be as great as 35°.

Belt drives

The drive to a conveyor is a special case of the familiar 'belt drive' which is used to transmit power from one shaft to another over a distance usually too great for a gear drive. A more expensive alternative is the chain drive, commonly seen on bicycles. (Can you suggest why belt drives are not used on bicycles?) In order to transmit power, belt drives depend on the friction acting between the belt and the pulleys or drums at each end of the drive. In operation, the belt will be subject to driving tensions F_1 and F_2 (Fig. 55).

F_1 is constant along the tight side of the belt, the belt being drawn on to the driving pulley and off the driven pulley by the driving pulley's rotation. F_2 will be less than F_1 by an amount equal to the frictional force acting between pulley and belt.

It can be shown that, for there to be no slip between the belt and pulley, the relationship between the belt tensions and the coefficient of friction is of exponential form,

$$\frac{F_1}{F_2} \leq e^{\mu \beta} \quad \ldots \ldots \ldots \ldots \ldots (9)$$

Fig. 54 A troughed belt conveyor in an enclosed gantry

Knowing the capacity of coal to be conveyed (and this must be the maximum load to be transported to the boilers for maximum possible power station output) and the average size of coal lump, the designer can calculate the width and speed of the belt. At fast speeds any slight wandering of the belt from side to side is difficult to correct even by careful alignment of the drums and idlers, and as a result maximum belt speeds are usually limited to 3 or 4 metres per second.

The maximum gradient against which the coal can be conveyed without rolling or sliding back is governed by the shape and size of the coal lumps and their condition (dry or wet), for the gradient depends on the coefficient of friction between the coal and the belt. The gradient may also depend on the belt speed and whether the feeding is continuous or intermittent. Without friction, of course, the belt would not move at all. This is true not only of belt conveyors but of all belt *drives*, where an endless belt is used solely to transmit power from one shaft to another, such as the fan belt in a motor car.

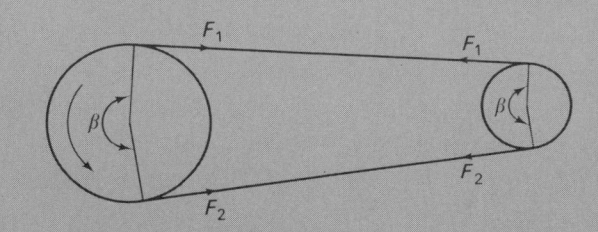

Fig. 55

where β is the angle of wrap of the belt on the pulley (see Fig. 55) and e is the base of natural logarithms

i.e. $$\mu\beta = \log_e F_1/F_2.$$

All belt drives exhibit a certain degree of slip between the belt and the pulleys, and chain drives must be used where positive transmission is required.

Also, the power transmitted by the belt from one pulley to the other is given by $(F_1 - F_2) v$ where v is the belt velocity.

If a conveyor belt is long and heavily loaded, the power needed to move it will be large, and the biggest possible value of F_1 is required, because the top, or tight strand has to pull the belt and load over all the idlers and also lift the load up an incline. From equation 9, we can see that F_1 can be increased by increasing the angle of wrap β, the coefficient of friction μ or the value of F_2, the tension of the lower strand of the belt. It is difficult to alter the value of μ, which is about 0.25 for a rubber belt on a steel drum. β can be increased by using a 'snub' drum as shown in Fig. 52, and F_2 can be increased by applying tension to the belt by means of a tensioner. Note that the action of the conveyor belt (and of course any belt drive) is independent of the *area* of contact of belt and pulley, although the greater the area the less the wear, and the width and thickness of the belt determine the magnitude of the tensile stresses set up in it. Also, a large contact area between the belt and the pulleys or drums will permit good dissipation of the heating which arises from slipping and hysteresis.

Friction clutch: elementary theory

Suppose the discs shown in Fig. 58, page 47, to be pressed together with an axial force P. This force acts normal to the surfaces in contact and gives rise to a friction force of magnitude μP, acting tangentially over the area of contact in the plane of the disc. If the ring of friction material is narrow, this frictional force can be taken as acting tangentially at the mean radius r, and since the moment of the frictional force is now $\mu P r$, the torque transmitted by the clutch is given by

Clutches

The driving drum of a belt conveyor is usually driven by an electric motor; in a power station this would be a 3-phase totally enclosed induction type and, depending on the belt width, load and speed, would have a power output of anything up to 100 kW. Figure 56 shows the layout of a simple drive system for a conveyor belt, where an electric motor drives the belt through a gearbox of the type shown in Fig. 25, page 21. The motor is connected to the gearbox input shaft by a coupling.

How can a belt be set in motion? When the motor is switched on, one of two things could happen. If the motor torque is great enough, the motor shaft will accelerate fairly rapidly, initially causing the driving drum to slip on the belt, and this may give rise to high rates of belt wear and heating. Also the sudden movement of the belt may cause it to whip or jerk, putting an undue strain on the belt material, as well as spilling its contents. On the other hand the torque developed by an induction motor at starting (when the slip is large: see *Electrical Fields and Devices*) is much lower than when running at speed, and this low starting-torque may not be enough to move the loaded belt, with the result that the motor stalls and burns out.

A similar problem occurs in a motor car. If the crankshaft of the car engine were directly connected to the rear wheels of the car the engine would not be able to develop enough power to start the car moving, even in low gear. There is also the problem of starting the engine in the first place, for it must be turned over fairly quickly if it is to be made

Fig. 56 Conveyor drive unit

$$T = \mu Pr.$$

The torque transmitted can be increased, therefore, by increasing (a) the mean radius r, (b) the pressure of the springs forcing the plates together, and (c) the coefficient of friction μ. If r is limited (by adjacent machinery for instance) the torque transmitted will depend on P and μ. μ is limited by the nature of the friction material used (usually from 0.3 → 0.5) and the maximum pressure by the stiffness of the springs used. (In car clutches what limits the spring stiffness?) For a given clutch-size the torque transmitted can, however, be increased by using a disc with friction material on both sides as shown in Fig. 57. In this case the torque transmitted is

$$T = 2\mu Pr$$

because there are two surfaces which give rise to frictional forces.

It is important to realise that the narrower the ring of friction material, the greater is the mean radius r (as long as the outer diameter of the friction ring remains constant). Hence maximum torque is transmitted when the ring is narrow. It must not be made too narrow, however, or the wear that occurs whilst the drive is taken up and the clutch is slipping will be excessive, and the resultant heating will be so great that the coefficient of friction of the surfaces may be reduced considerably. Note that the torque does *not* depend on the *area* of contact of the surfaces, although a compromise must be reached between using a narrow ring which would wear rapidly and become overheated, or a wide friction ring giving rise to little wear but needing a larger overall diameter.

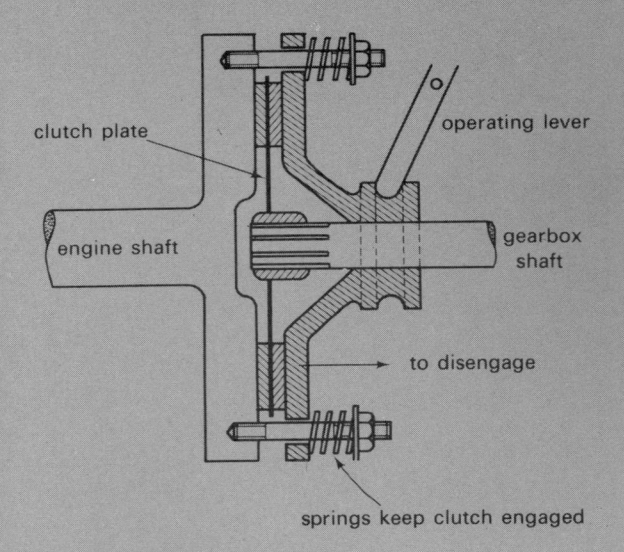

Fig. 57 Simplified section of car-type clutch

to 'fire', and this would be impossible, unless, of course, there were always people handy to give a push start! Also, every time the car stopped the engine would stall, and gear changing would be anything but quiet! Obviously, we need to be able to disengage the engine output shaft from the gearbox input shaft, and there is also the problem of taking up the drive gradually so that the torque transmitted to the driving wheel can be increased slowly, preventing the engine from stalling.

A number of devices are available which carry out both of these functions. Of the many different types the commonest is the friction clutch, and the name itself implies that it relies on friction for its operation. In its simplest form a friction clutch consists of two discs A and B each fixed to shafts aligned along the same axis (see Fig. 58). The revolving driving disc A, which has a smooth face, is required to transmit power to the stationary disc B, which is faced with a ring of high-friction material as shown. If the discs are pressed together endways the faces come into contact and a friction force will act, tending to slow down disc A. As the axial force is increased the frictional force will become great enough to cause the driven disc B to move. B will gradually speed up until it is ultimately revolving at the same speed as A, and the clutch is said to be fully engaged, transmitting power from the driving to the driven shaft, just as if there were a rigid coupling joining the shafts. As the drive begins to be taken up, there will be a high degree of slip between the discs, and a small amount of wear (depending on the quality of the friction facing). When fully engaged there will of course be zero slip and no wear. Car clutches remain in the engaged position until a gear change is required. At this point the clutch is disengaged and the

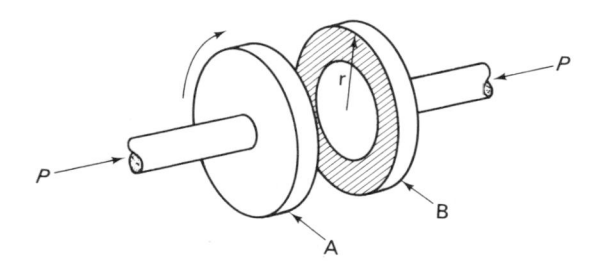

Fig. 58

The action of a simple external brake

One of the oldest and simplest brakes is the *band* or rope brake, as illustrated in Fig. 59. It depends entirely on the frictional force acting between the wheel and the rope or band. Suppose the brake is applied to a shaft rotating as shown. The torque causing the wheel to rotate must equal the torque provided by the friction force acting at radius r. The torque developed by the braking action is given by

$$T = (F_2 - F_1)r$$

and if the brake torque is greater than the driving torque the wheel will eventually stop, and the brake will prevent it from further rotation. The rope brake is commonly used to measure the power output of engines and electric motors, because the tension in the rope can easily be measured by means of spring balances. Figure 60 shows the band slipping on the wheel. If the spring balance reading is S, the torque on the shaft is given by

$$T = (S - mg)r$$

where $(S - mg)$ is the friction force acting at the rope/drum interface. If the angular speed N of the shaft is measured in revolutions per second, the power P is found from

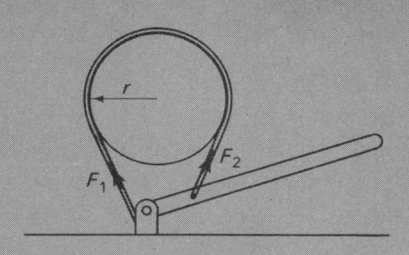

$$P = T\omega = T2\pi N.$$

Fig. 59 Band brake

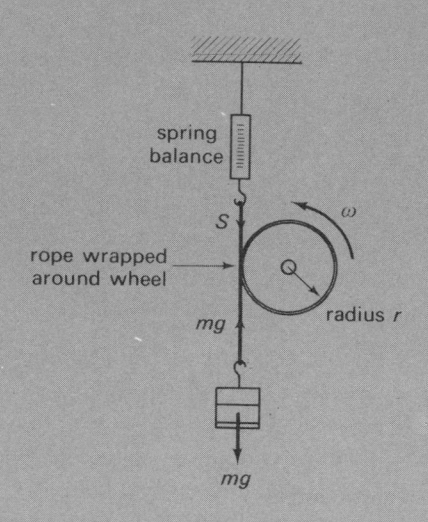

Fig. 60 Rope brake

drive is taken up. Springs are used to apply the axial pressure. To increase the capacity to transmit power, a disc with friction material on both sides is used. Figure 57, page 46, shows a simplified diagram of a car-type clutch.

In the drive from an electric motor to a conveyor belt it is often inconvenient to install a friction clutch (can you suggest why?). It is easier to fit extra starter resistances into the rotor circuit which increase the rotor torque when the slip between rotor and stator is high, i.e. at starting. To prevent the torque from dropping subsequently, these resistances are cut out as the motor accelerates. If it is still necessary to have some sort of device to prevent jerking of the belt, or slipping between belt and drum, then a *fluid coupling* can be installed. This depends for its action on the change in kinetic energy of a fluid moving in special passages in adjacent 'rotors', one on the driving shaft, the other on the driven shaft. For a full description of the fluid coupling you are advised to refer to any good book on the design of machine parts.

Brakes

Suppose it is required to stop the conveyor belt shown in Fig. 54. It is not good enough to simply switch off the supply to the electric motor, as the weight of coal on the inclined belt will cause it to run back, covering the feed device with coal. To prevent this happening a brake is fitted between the electric motor and the belt-driving drum. (Should the brake be fitted on the gearbox input shaft, or on its output shaft, and why?) The brake operates at the moment that the electric circuit to the motor is broken; this is usually effected by a 'trip' mechanism, whereby breaking the circuit to the electric motor also breaks the circuit to a solenoid (see Fig. 62).

When the solenoid N is de-energised, the spring S holds the brake shoes M, lined with high-friction material, against the rim of the brake wheel W. The wheel W slows down under the action of the friction force at the shoe-wheel interface, until it stops. Spring S provides a force great enough to prevent further turning of the wheel. Obviously the brake torque

The action of a simple drum brake

Figure 61 shows two arrangements of an internal expanding, or drum brake. A force W is applied by either mechanical or hydraulic means as described in text A, and this gives rise to a reaction R acting on the shoe, which produces a frictional force μR acting on the shoe in the direction shown (Fig. 61a).

The shoe is in equilibrium; taking moments about the brake shoe pivot:

$$Wa = Rb + \mu Rr$$

hence
$$R = \frac{Wa}{(b + \mu r)}.$$

Now the braking torque acting on the drum is due to

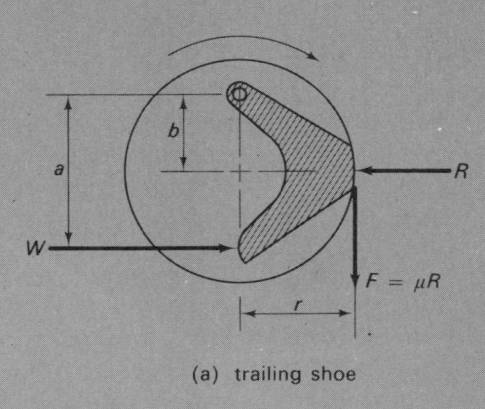

(a) trailing shoe

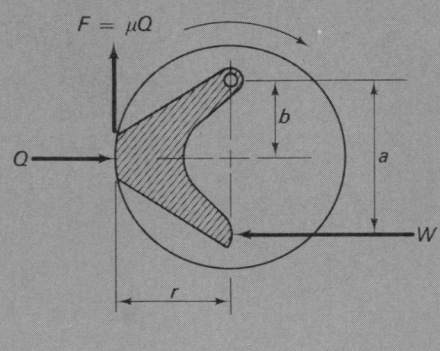

(b) leading shoe

Fig. 61

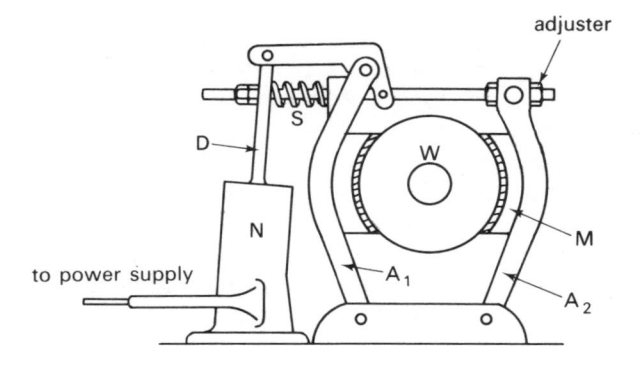

Fig. 62 Rim brake

provided by the spring and the linkage must be great enough to balance the torque tending to cause the belt to run back. If the electric motor is switched on, the solenoid is energised, the rod D moves downwards, and the linkage forces the tops of the brake arms A_1 and A_2 apart, compressing spring S and releasing the brake shoes from the wheel rim. The greater the coefficient of friction between the braking surfaces, the smaller is the overall size of the brake to resist a given torque.

Internal expanding brakes

The type of brake discussed above is called a *rim* brake, or an *external contracting* brake. It suffers from a number of disadvantages, mainly that it is exposed to dirt and abrasive dust unless covered over, and it is not very compact. These are obvious disadvantages as far as car brakes are concerned, and nearly all car brakes nowadays are of the internal expanding (or drum) type, or alternatively the disc type which will be considered in the next section.

Figure 63 shows a simple car drum brake where both brake shoes share a common anchor pin. Normally a small clearance exists between the shoes and the inside surface of the brake drum. When the brake pedal is applied, the pressure at the pedal is transmitted via the hydraulic brake fluid from the master cylinder to small pistons at the free ends of the shoes. The pressure on these pistons forces the shoes against the brake drum, and the friction force acting at the brake drum radius creates a torque in opposition to the motion

the frictional force μR acting at a radius r, hence the brake torque T_1 is given by

$$T_1 = \frac{\mu War}{(b + \mu r)} \quad \cdots \cdots \cdots \cdots (10)$$

You may think at first sight that the brake torque for the shoe shown in Fig. 61b would be exactly the same, but consider the equilibrium of the shoe under the action of the forces shown. Taking moments about the shoe pivot

$$Wa = Qb - \mu Qr$$

hence
$$Q = \frac{Wa}{(b - \mu r)}$$

and the brake torque

$$T_2 = \mu Qr$$

$$T_2 = \frac{\mu War}{(b - \mu r)} \quad \cdots \cdots \cdots \cdots (11)$$

Because of the sign difference in the denominator, T_2 is *greater* than T_1 for the same actuating force W.

The shoe that gives rise to T_2 is called the leading shoe, the other shoe is the trailing shoe.

To give an idea of the magnitudes of the torques involved, consider a 0.20 m diameter drum brake in which a = 0.16 m, W = 1 000 N, and the coefficient of friction μ = 0.5.

$$T_1 = \frac{\mu War}{b + \mu r} = \frac{0.5 \times 1\,000 \times 0.16 \times 0.10}{0.16 + (0.5 \times 0.10)} = 38 \text{ N m}$$

$$T_2 = \frac{\mu War}{b - \mu r} = \frac{0.5 \times 1\,000 \times 0.16 \times 0.10}{0.16 - (0.5 \times 0.10)} = 73 \text{ N m}.$$

Hence in this case the braking torque T_2 developed by the leading shoe is almost twice that produced by the trailing shoe.

You may like to consider equations 10 and 11 with a view to finding out how the designer of a drum brake can best increase the braking torque. You should bear in mind that the theory outlined above has been simplified by the fact that the area of contact between shoe and drum is small. A more rigid theoretical treatment for the brake shown in Fig. 63 would involve integrating the moment of the friction force on a small element of the friction material over the arc of contact between the shoe and the drum.

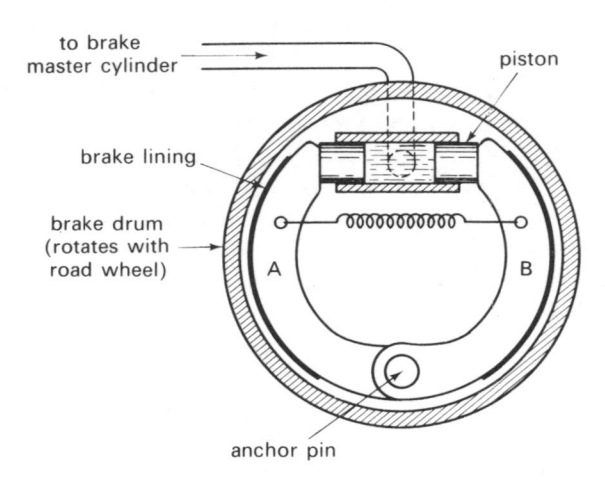

Fig. 63 Car-type drum brake

of the wheel, causing the wheel to slow down and stop. When the brakes are applied, both shoes move into contact with the drum, but shoe B tends to follow the rotating drum and wedges itself between the drum and the anchor pin. Large reaction forces are set up, giving rise to high friction forces, and if the correct friction lining material is not used the brake may even be too sensitive. (The friction assists in applying the braking torque; this is a good example of 'positive feedback'. See *Electronics, Systems and Analogues*.) B is said to be a leading shoe, and A a trailing shoe; usually the leading shoe contributes well over 50 per cent of the total braking torque on the drum. Obviously if the direction of drum rotation is reversed, A becomes the leading shoe and B the trailing shoe. You may care to investigate designs of drum brakes in which *both* shoes are leading shoes.

Disc brakes

Car brakes do not only have to apply large braking torques to the brake drums for short periods of time for stopping purposes; they also have to dissipate large quantities of heat energy when, say, the car descends an incline. The change in kinetic energy of the car when braking, or the change in potential energy when it descends an incline, must be converted into heat energy at the brake surfaces. The rate of heat dissipation is about the same (10 kW) for a car braking hard on a level road from 40 miles per hour as for the brake being used to maintain a steady 50 miles per hour down a 1 in 30 slope. In both cases, however,

Disc brakes

Let the pressure of the fluid in the stationary calliper of the disc brake shown in Fig. 64 be p, and the area of each piston be a. The force on each piston will be given by the product pa.

This is the normal force acting between each pad and the steel disc, and will give rise to a friction force equal to μpa. Thus the total friction force due to both pads is equal to $2\mu pa$, and because this acts at a radius r, say, the total braking torque is given by

$$T = 2\mu par.$$

Note that a is the area of each piston, *not* the pad area. Once again the pad area does not affect the braking torque, but does influence the wear of the pads and their temperature-rise during braking.

Friction in screw threads

Rotation of the screwed shaft shown in Fig. 65 to raise the load mg, means in effect raising mg by means of a plane inclined to the horizontal at an angle α, i.e. the helix angle.

Friction opposes the motion of mg up the plane, and the resultant X of the friction force μR and the normal force R is shown in Fig. 65b. X will be at the angle of friction ϕ from R when the horizontal force P is just enough to overcome the friction and cause the load to move. From the triangle of forces, the horizontal force P is therefore

$$P = W \tan(\alpha + \phi).$$

It can be shown mathematically that,

$$\tan(\alpha + \phi) = \frac{\tan \alpha + \tan \phi}{1 - \tan \alpha \tan \phi}$$

$$\therefore P = W\left[\frac{\tan \alpha + \tan \phi}{1 - \tan \alpha \tan \phi}\right].$$

Now $\tan \alpha = \dfrac{\text{lead of screw}}{\text{thread circumference}} = \dfrac{\ell}{2\pi r}$

and $\tan \phi = \mu$ (see eqn. 5, text B, page 10)

hence $$P = \frac{W\left(\dfrac{\ell}{2\pi r} + \mu\right)}{1 - \dfrac{\ell \mu}{2\pi r}}$$

$$P = W \frac{\ell + 2\pi r \mu}{2\pi r - \mu \ell}.$$

Applying a horizontal force H at the end of a lever of length y is comparable to applying the force P at the

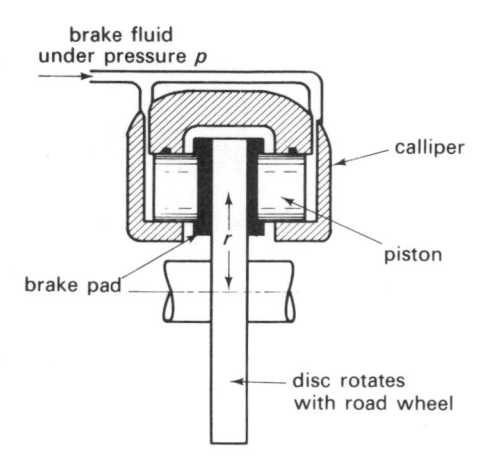

Fig. 64 Section through an hydraulically operated disc brake

the engine will tend to act as a brake if the car is in gear, and these figures would require modification if this is taken into account. Drum brakes do not provide for efficient cooling either of the brake shoes or of the drums, and at the high temperatures developed, the coefficient of friction between the shoes and drum may drop considerably. The problems of heating can be partly reduced by using the disc brake, in which a heavy steel disc attached to the wheel passes between the legs of a stationary calliper (see Fig. 64). This holds two actuating cylinders fitted with pistons. The outer ends of the pistons are fixed to friction pads, which are simultaneously forced against the disc when the brakes are applied. Heat dissipation rates are much higher for disc brakes than for drum brakes because of the rapid flow of cool air across the disc.

Fasteners

Nearly all fastening devices are basically friction devices although on a much less obvious scale than devices like brakes and clutches. Many fasteners incorporate screw threads, where the friction acting between the faces of adjacent threads prevents the fastener from loosening. Even adhesives depend on friction in that they rely on strong chemical bonding processes, not only between the materials to be joined and the adhesive, but between the molecules of the adhesive itself. Many types of fasteners are discussed in greater detail in *The Use of Materials*.

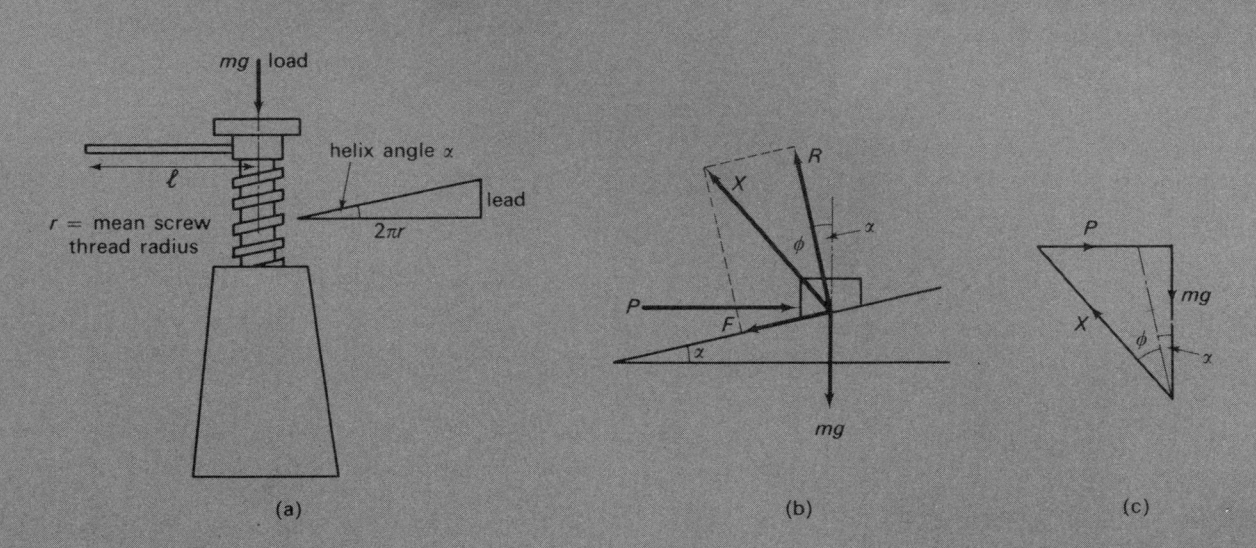

Fig. 65 The screw jack

screw thread to slide the load up the plane (i.e. up the screw thread). Therefore the torque produced by the horizontal force H required at the end of the lever, of length y is given by

$$Hy = Pr$$

$$Hy = \frac{Wr(\ell + 2\pi r\mu)}{2\pi r - \mu\ell}.$$

Now consider movement of the load *down* the plane, i.e. the unwinding of the jack. This is the same as applying P' to a load W on the inclined plane as shown in Fig. 67.

In this case

$$P' = W \tan(\phi - \alpha)$$

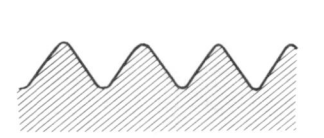

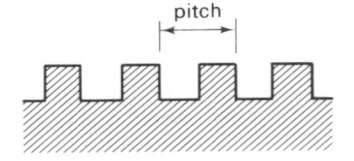

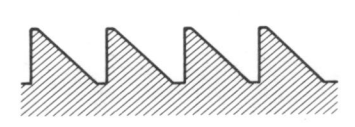

(a) Vee thread   (b) square thread   (c) buttress thread

Fig. 66 Screw thread profiles

The screw thread was one of the most significant inventions in the history of engineering, for it has provided a means of holding together removable parts on machines, clamping together structural members, and transmitting power. Screw threads perform a variety of other functions such as providing for the fine adjustment of measuring instruments.

Essentially there are two forms of screw thread: the vee thread and the square thread (see Fig. 66), although the precise dimensions vary throughout the world and screws are not yet internationally interchangeable. Vee threads provide greater frictional forces than square threads, hence the latter are used where slipping or movement is required. Generally, vee threads are used in nuts, bolts and screws, whereas the square thread is used for transmitting power, e.g. lead screws on lathes, screw jacks.

Special threads have been developed; one example is the buttress thread used on the standard type of woodworkers' quick-release bench. You may care to take a look at this vice mechanism to see why the buttress thread is used.

In the nut and bolt, which usually incorporate a vee thread, the load is taken on one side of the thread only as shown in Fig. 68a. As the nut is tightened the bolt is stretched, i.e. a tensile stress is set up along its length, usually less than the yield stress. This tensile stress gives rise to very high forces normal to the sides of the threads, which produce correspondingly large friction forces, preventing the nut from unscrewing. A spanner must be used to overcome the friction forces.

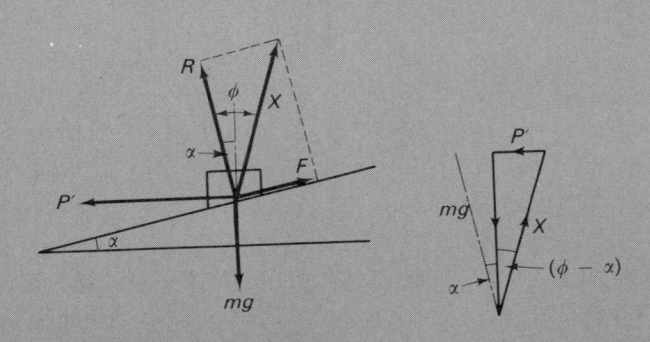

Fig. 67 Unwinding the jack

and the torque to unwind the jack (Hy) can be found in the same manner as before.

If ϕ is greater than α, as shown, a horizontal force P' from left to right will be required to unwind the jack, but if ϕ is *less* than α, X will lie to the left of the vertical, and a horizontal force will be required in the opposite direction to P' to hold the load on the plane. In other words unless this force is applied the load will slide down the plane (if ϕ is less than α) and this of course means that the jack will unwind itself under the influence of the load it supports — this is obviously undesirable as far as a car jack is concerned!

Efficiency of screw threads

The screw jack is a simple machine whose efficiency η, is given by

$$\eta = \frac{\text{useful work done in raising load}}{\text{work done in turning screw thread}}$$

$$= \frac{\text{output work}}{\text{input work}}.$$

With no friction acting, the screw jack (or indeed any screw thread) would be 100% efficient, i.e. all the work done by the force at the end of the lever would be converted into increasing the potential energy of the load. In practice the effect of friction is to convert some of the input work into useless heating at the screw threads in contact. Let us call this the wasted work.

The input work = gain in p.e. + wasted work

and $$\text{efficiency} = \frac{\text{gain in p.e.}}{\text{gain in p.e.} + \text{wasted work}}.$$

When the load has been raised, the potential energy gained is available to lower the load. When the load is lowered through the same height the same amount of wasted work will be done. If the potential energy is greater than the wasted work, there is enough energy available to cause the jack to unwind under the

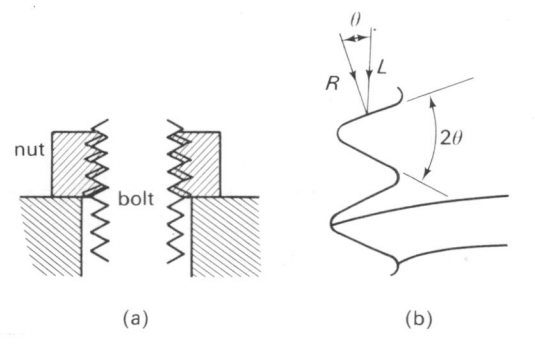

Fig. 68

For a given load the frictional forces between vee threads are greater than those between square threads: this is easily seen from Fig. 68b. The vertical load L is equal to the vertical component of the normal reaction R

i.e. $\quad\quad\quad L = R \cos \theta$

∴ $\quad\quad\quad R = L/\cos \theta \quad$ (vee thread)

compared with $\quad R = L \quad$ (square thread).

Hence the normal force for a vee thread is greater than that for a comparable square thread, and the friction force ($= \mu R$) is correspondingly greater.

Any screw thread can be thought of as being a special case of the inclined plane; the problem is not dissimilar to that of coal on an inclined conveyor belt. Figure 65 shows a load on a square-threaded screw jack, such as those used to raise cars. The thread is simply an inclined plane 'wrapped' around a cylinder in the form of a *helix*. The *helix angle* α is, of course, the angle of inclination of the inclined plane. For every rotation of the thread the load will be raised through a distance called the *lead*.

The power screw is a threaded shaft which, when turned, is required to transmit power to a device such as the saddle on a lathe.

A screw-type car jack is used to raise a load by turning a threaded vertical shaft. It is possible, however, for the thread to have a helix angle such that once the load is raised, it promptly descends under its own weight, 'unwinding' the jack! It is shown below that the efficiency of the screw (the potential energy gained by the load as it is raised, divided by the work done to raise the load) must be less than 50 per cent if the jack is not to 'unwind'. On the other hand the efficiency must not be so low that power is wasted unnecessarily in raising the load.

influence of the load. If the wasted work is equal to the potential energy, the load will cause the jack to unwind with no gain in kinetic energy, and the efficiency will be

$$\eta = \frac{\text{gain in p.e.}}{\text{gain in p.e.} + \text{gain in p.e.}} = \frac{1}{2}.$$

To sum up, if the efficiency is greater than 0.5, the load accelerates downwards, unwinding the jack; if the efficiency equals 0.5, the load slowly descends; if the efficiency is less than 0.5 the load remains stationary after it has been raised. In the last case the screw is a non-return mechanism. For obvious safety reasons the car-jack must have an efficiency of less than 0.5.

If the load just falls under its own weight the tangent of the angle of the thread (or inclined plane) is μ. Any screw mechanism, whose lead/circumference ratio is less than the coefficient of friction, will be non-returning.

Jamming

Consider the movable arm A of the clamp shown in Fig. 70c. The axial force on the screw thread as the screw is tightened tends to slide arm A to the left along bar B. If friction acting at the sliding surfaces did not cause the slide and bar to jam together, the clamp as shown would be useless. On the other hand many sliding parts of machines must be designed so that they do not stick or jam under the application of applied forces. This is especially true of machine tools, for example in a lathe where a 'saddle' driven by a power screw has to move along slides without jamming (see Fig. 18, text A, page 16).

Jamming of sliding parts is not only encountered in machines. Anyone who has tried to close a sash window (or a drawer in a chest of drawers) in a hurry will have come across the problem of jamming, and it would be thought of as being the penance for trying to do something too quickly! The more hurried the action, the longer it seems to take!

How do these jamming forces arise, and how can the designer of, say, a chest of drawers or the slideways of a lathe know whether jamming is likely or not?

Consider the plan view of a sliding block ABCD shown in Fig. 69a. An applied force acts on the block along its centre line and parallel to the sides of the guide. Friction acts over the bottom surface of the block and if P is greater than $\mu m g$ the block will slide easily. (It is assumed that because of the necessary small clearance between the block and the sides of the guide there are no friction forces along the edges AB and CD.) Jamming occurs when the force P is

Clamps and cotters

Many devices are used in the production engineering industries to locate and hold pieces of material whilst they are machined. The tool that cuts the metal is also usually held in some sort of chuck, grip or clamp. A familiar example is an electric drill: the drill bit is clamped tightly in the chuck and the work piece is gripped in a vice; the friction acts between the drill bit and chuck and between the work and the vice jaws to prevent slipping.

Figure 70 shows a number of clamping devices in which the large reaction forces between the surfaces in contact give rise to the high friction forces necessary to prevent sliding.

Cotters are tapered bars or pins, used to join shafts or structural ties in tension. The force required to drive in the cotter depends upon the tension, the coefficient of friction at the adjacent surfaces and the taper angle. The advantage in using tapers to clamp parts together is that a small axial force can produce large friction forces at the mating taper surfaces; also, tapered pins can be driven in to take up any slack between adjacent parts, and are easily released. For these reasons tapered cotter pins are used to fix the pedal cranks on bicycles. In machine tools high torques are often transmitted (or resisted) by frictional forces acting on a taper. One example is the taper shank drill bit where the axial force causing the drill to 'bite' into the material creates high frictional forces on the taper surfaces, which effectively lock the drill bit to the chuck. Another example of friction associated with tapers is the wedge that holds open a door.

This chapter should have given you an idea of how important a part friction plays in devices essential to the operation of machines. Against this usefulness must be balanced the wear that the friction causes, e.g. the wear of brake and clutch linings, with the necessity for periodic inspection and renewal. It would be ideal if friction devices could operate satisfactorily whilst at the same time giving no wear. Much research needs to be carried out with a view to developing cheap high-friction surfaces that show minimal wear and require no maintenance.

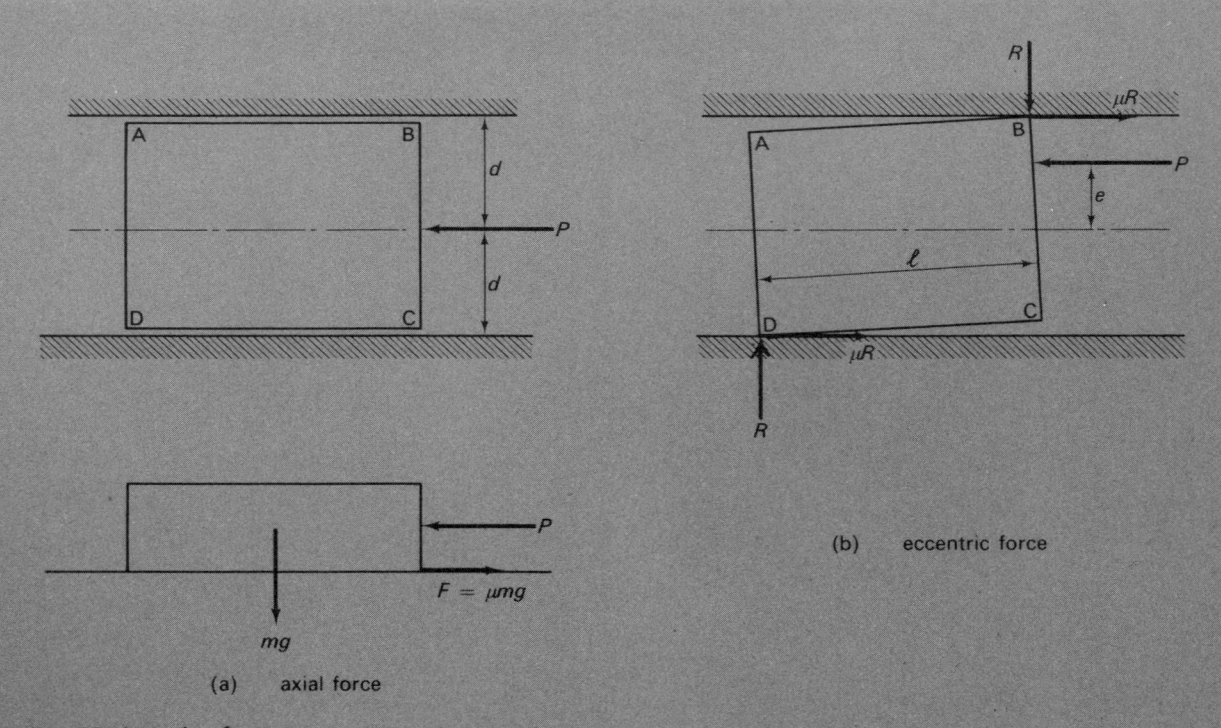

Fig. 69 Jamming forces

(a) axial force

(b) eccentric force

applied *eccentrically*, i.e. at a distance e from the centre line. Due to the small clearance between the block and the guide and the eccentricity of the applied force, the slide will tilt as shown and friction forces will act at the opposite corners which are in contact with the guide. Remember that the friction forces tend to oppose the motion of the block.

The resistance to motion due to friction at the

(a) 3-jaw lathe chuck

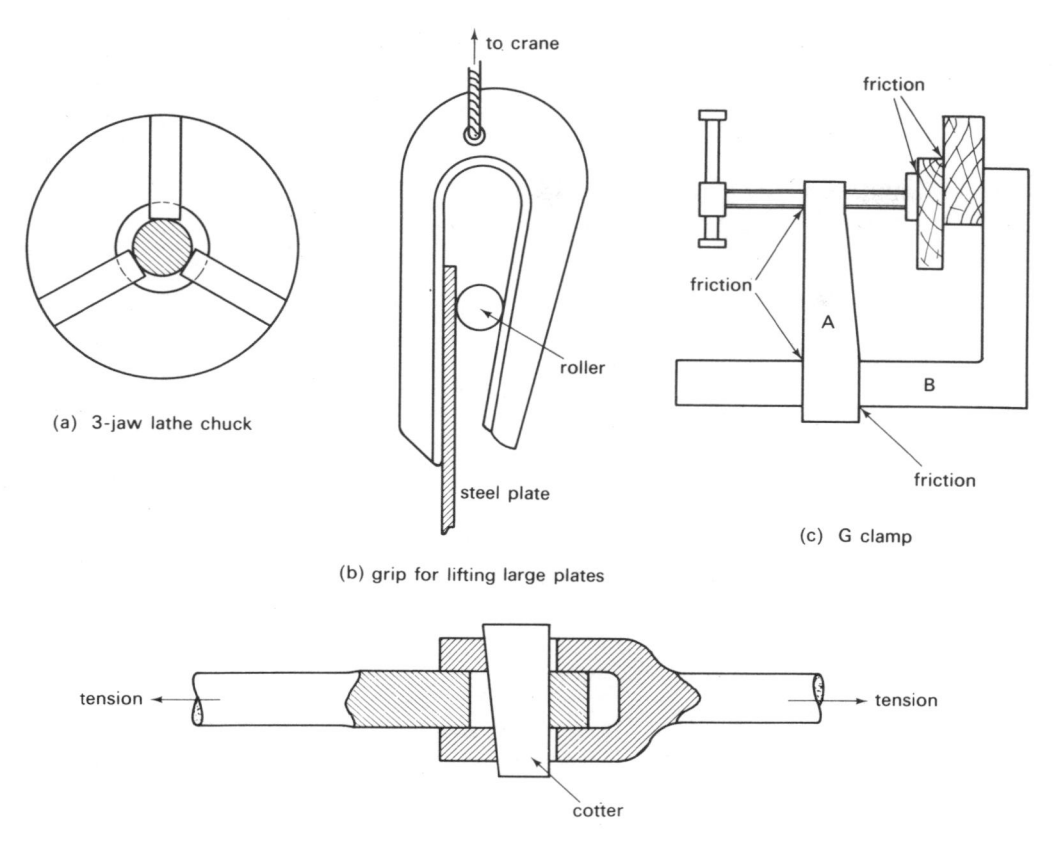

(b) grip for lifting large plates

(c) G clamp

(d) method of joining shafts in tension

Fig. 70 *Some clamping devices*

sides of the block = $2\mu R$, also the resistance due to friction acting on the lower surface of the block = μmg.

Equating the forces from left to right of Fig. 69 when the slide is just on the point of moving,

$$P = \mu mg + 2\mu R \quad \ldots \ldots \ldots \ldots (12)$$

And taking moments about a suitable point X on the centre line

$$eP + d\mu R = R\ell + d\mu R$$
$$eP = R\ell$$
$$R = \frac{Pe}{\ell}.$$

Substituting in equation 12

$$P = \mu mg + 2\mu \frac{Pe}{\ell}$$
$$P\left(1 - 2\mu \frac{e}{\ell}\right) = \mu mg$$
$$P = \frac{\mu mg}{1 - 2\mu \frac{e}{\ell}} \quad \ldots \ldots \ldots (13)$$

If $e = \frac{\ell}{2\mu}$,

the denominator of equation 13 is zero, and the force P to cause sliding becomes infinite, meaning that the block jams.

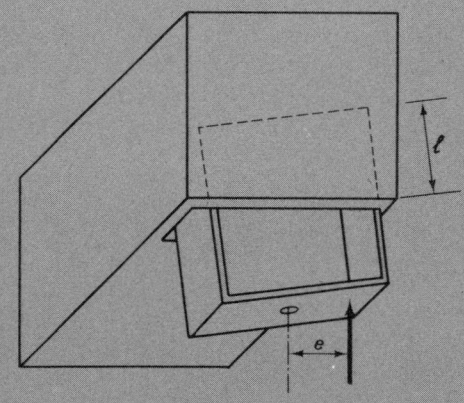

Fig. 71

Hence if

$$e \geqslant \frac{\ell}{2\mu} \quad \ldots \ldots \ldots \ldots (14)$$

the block will jam in the guide.

(The same result is true if the frictional force μmg acting under the block is neglected in the analysis.)

You may like to investigate how far this relationship holds true for a drawer as shown in Fig. 71. Try varying the dimension and measuring e for jamming just to occur. Many clamping systems could be made on the basis of equation 14, but it is sounder engineering practice to incorporate a more positive form of locking device, e.g. a screw or peg might be used to hold the arm of the clamp shown in Fig. 70c to the bar B.

Indexes

Index

When following up a particular reference you are advised to examine also the pages immediately following those indicated in the Index. The letters A and B following each page number refer to the upper and lower texts respectively.

Air   42B
Alcohols   20B
Angle of friction   10B
Angle of wrap   45B
Animal oils and fats   21B
Annular groove seals   26A
Aquaplaning   43B
Area of contact   6B
Ash disposal   9B
Ash handling   8B

Babbit metals   12A
Ball bearings   15A,B
Band brakes   47B
Bearing materials   12A,B
Bearings   Chapter 2, 38B, 41B
Bearing seals   26A
Belt drives   44B
Bonding forces   25B
Boundary lubrication   40B
Brakes   47B, 48A
Brasses   12A
Brunell hardness   32B
Bubbles   26B
Bushes   12A
Buttress thread   52A

Capillary action   28B
Capillary viscometer   36A
Carbon   19B, 41A
Centre-line average (CLA)   30B
Chloroethane   20B
Chlorohexane   20B
Choice of bearings   17A
Choice of lubricants   40A
Chucks   54A
Clamps   54A
Clutches   45B, 46A
Coal-fired power stations   2A,B
Coal handling   5A
Coefficient of friction   9B, 17A
Concentric oil bearing   35B
Consistency of greases   40A
Conveyor belt   43B, 44A
Cooling   25A, 39A
Costs   10A
Cotters   54A
Cup and cone bearing   15A

Degradation of oil   38A
Design criteria   10A
Dichloroethane   20B
Dimethyl hexane   19B
Disc brakes   50A,B
Drip-feed lubricators   19A
Drum brakes   48B, 49A
Dynamic viscosity   34B

Efficiency   52B
Efflux of viscometers   36A
Electrical power demand   1B
Electrical power generation   2A,B
Ethane   19B
Ethylene glycol   20B
Excess bubble pressure   26B
Extreme pressure (EP) lubricants   40B

Fan blade wear   7A
Fasteners   51A
Fatigue   12B, 15A
Felt seals   28A
Fluid shear   33B
Foaming   38A
Force-feed oil systems   19A
Friction   2A, 5B
   laws of   6B

Gas lubrication   42B
G-clamps   55A
Gears   20A
Glycerol   22B
Graphite   25B, 41A
Gravity-feed oil systems   19A
Grease cups   23A
Grease lubrication   22A,B
Grease nipples   23A

Hardness   31B
Heat   18B, 25A, 39A, 50A
Helix   53A
Hexane   19B
Hydraulic brakes   50A
Hydrocarbons   19B
Hydrodynamic lubrication   38B
Hydrostatic lubrication   40B
Hysteresis loss   14B, 21A

Inclined plane   43B
Interfacial forces   25B
Intermolecular forces   18B

Jamming   53B
Journal bearings   11A

Kinematic viscosity   35A,B
Kinetic friction   8B

Labyrinth seals   26A
Lag   30A
Lathe chuck   55A
Lead   51B, 53A
Leading shoe   49B, 50A
Limiting friction   5B
Linear bearings   15A
Lip seals   28A
Lubricants   18B, Chapter 5
Lubricating oils   19B
Lubrication   11B, Chapter 3

Machine tools   15B
Maintenance   9A
Measurement of hardness   32B
Measurement of surface tension   26B
Measurement of viscosity   36A
Mechanical seals   29A
Methane   19B
Mica   32A
Michell bearing   13B
Molybdenum disulphide   23B, 40B, 41A

Newton   33B
Non-return mechanisms   53B
Nuclear power stations   2A
Nylon   41A

Octane   19B
Oil-fired power stations   2A
Oil lubricators   16A
Oil wedge   13B
Oxidation   10B, 38A

Paraffins   21B
Petroff   35B
Pitch   52A
Plain bearings   11A
Porous bearings   13A
Power transmission   45B
PTFE   9B, 25B, 41A
Pulverising mill   4A

Redwood viscometer   37A
Reynolds   38B
Rim brakes   48A
Ring oiler   20A
Roller bearings   14A,B
Rolling friction   13B
Rope brake   47B
Roughness   30A,B
Running-in   32A

Screw threads   50B
Searle's apparatus   29B
Self-aligning bearings   15B
Shear stress   7B, 34B
Shims   12A
Shore scleroscope   32B
Sintering   13B
Sleeve bearings   11A
Slideways   15A
Sliding fraction   8B
Slinger seals   26A
Soaps   22B
Sodium stearate   22B
Splash lubrication   21A
Spray lubrication   22A
Square threads   52A
Starting torque   46A
Static friction   5B
Stauffer cup   23A
Stearic acid   22B
Stearing   22B
Surface adhesion   16B, 23B
Surface energy   27B
Surface roughness   6B
Surfaces   Chapter 4
Surface tension   16B, 23B, 25B
Syphon lubricators   19A

Talc   25B
Thrust bearings   11A, 13B
Tilting pad bearing   13B
Trailing shoe   49B, 50A
Tribology   1A
Trichloroethane   20B
Turbine bearing   13A
Turbine lubrication   24
Tyres   43B

Van der Waals forces   25B
Vee threads   52A
Vegetable oils and fats   21B
Velocity gradient   34B
Vickers hardness   32B
Viscometers   36A
Viscosity   33B, 34A
Viscosity meter   37B

Waste utilisation   6A
Water   21B, 39A
Wave   25B
Waviness   30A, B
Wear processes   11B
Wear rate   6A
Wear resistant materials   9A
White metal   12B
Worm gear   21A

Yield stress   7B

Examination index

Only the first page of each relevant section of the text is listed against each heading. The letters A and B following each page number refer to the upper and lower texts respectively.

Not all the sections or treatments listed are equally relevant to all syllabuses. When working for the final examination make sure that the material you are revising is included in your syllabus.

Materials science

Surface energy   27B
Surface tension   25B
Viscosity   33B, 34A

Mechanics

Brakes   47B
Friction   2A, 5B
Inclined plane   43B
Machine efficiency   52B
Power transmission   45B
Rolling friction   13B